Posthuman Studies

Stefan Lorenz Sorgner (ed.)

Volume 3

Russell Blackford

At the Dawn of a Great Transition

The Question of Radical Enhancement

Schwabe Verlag

MIX
Papier aus verantwortungsvollen Quellen
FSC® C083411

Bibliographic information published by the Deutsche Nationalbibliothek
The Deutsche Nationalbibliothek lists this publication in the Deutsche Nationalbibliografie; detailed bibliographic data are available on the Internet at http://dnb.dnb.de.

Cover illustration: © Jaime del Val
Cover: icona basel gmbh, Basel
Graphic design: icona basel gmbh, Basel
Typesetting: 3w+p, Rimpar
Print: CPI books GmbH, Leck
Printed in Germany
ISBN Hardcover 978-3-7965-4189-6
ISBN eBook (PDF) 978-3-7965-4214-5
DOI 10.24894/978-3-7965-4214-5
The ebook has identical page numbers to the print edition (first printing) and supports full-text search. Furthermore, the table of contents is linked to the headings.

rights@schwabe.ch
www.schwabe.ch

For Nick Agar

Contents

Acknowledgements

I've dedicated this book to Nick Agar, a valued colleague who has always been kind and supportive – and, many years ago now, one of the examiners of my Monash University doctoral dissertation. At a crucial time in my life, Nick encouraged me to go on with the challenging, important task of philosophical bioethics. In the following pages, I spend much time, energy, and ink discussing, and often disputing, his published views on radical enhancement, but none of that is meant to detract from the value of his work. Indeed, much of it has persuaded or influenced me even against my initial intuitions. I look forward to his responses to the arguments herein, as that is how intellectual progress is made.

Over the years, I've been fortunate to have mentors and friends who've helped me develop and clarify my thoughts on philosophical issues, not least those relating to emerging technologies and the prospect of human enhancement. On this occasion, I especially thank Damien Broderick, Justin Oakley, and James Hughes.

Thanks also to Stefan Lorenz Sorgner and Christian Barth – my series editor and commissioning editor respectively – who were instrumental in the publication of this book. Gregory Benford has always been helpful, and in this case I took inspiration from his thoughts on extraterrestrial aliens. (The thing is, they're *alien.*) A special shout out goes to Marcelo Rinesi, who helped me keep my head (somewhat) above water for many years by absorbing much of the work at the *Journal of Evolution and Technology.*

As always, Jenny Blackford has my heartfelt thanks for her support in numerous roles: from computer wrangler, to night-and-day sounding board, to first-response editor and much more. Jenny is a constant source of love, stability, and good sense in my life, and somehow (with remarkably few wistful sighs) she endures my physical and mental absences when I'm absorbed in writing volumes like this one.

All of the above are hereby absolved of any responsibility for whatever errors, misunderstandings, and solecisms remain herein despite my best efforts. I know from harsh experience that some always sneak through to publication. I'm quite capable of introducing them for myself, and I'm grateful for the help I receive in tracking down and eliminating as many as possible.

Thing about aliens is, they're alien.
– Gregory Benford, "Doing Alien"

Chapter One : The Question of Radical Enhancement

1. Into the unknowable

The thing about radical enhancement is, it's *radical.*

For comparison, Gregory Benford, the award-winning science fiction writer and distinguished astrophysicist, has often emphasized the sheer alienness of intelligent extraterrestrials, if such exist. If we ever encounter them, we must not assume that their thought processes and motivations make intuitive sense in human terms. This idea appears in one of Benford's early short stories, "In Alien Flesh" (first published 1979; collected in Benford 1986), whose protagonist has worked – in terrifying circumstances – studying the alien Drogheda: huge marine creatures, comparable to whales, whose impressive intellects are known only via their mathematics, since they have not built a technological civilization. In an afterword to the story, Benford discusses the frequent representation of aliens in science fiction either as monsters or as analogues for indigenous tribes, but for him what most matters about aliens is that they would be very strange by our standards: "[T]he most compelling aspect of aliens is their fundamental unknowability" (Benford 1986, 31).

Benford makes the point succinctly in another story, "Doing Alien", first published in 1994: "Thing about aliens is, they're alien" (Benford 2000, 41, 43). This phrasing, or similar, is now closely associated with Benford. A variation – "The thing about aliens is, they're alien" – appears in his story "A Hunger for the Infinite" (Benford 1999, 376), where the point is that *we*, in the story's particular circumstances, are alien.

We might hope that radically transformed human beings, their capacities augmented through technological means such as genetic engineering, brain–machine interfaces, advanced prosthetics, powerful pharmaceuticals, or extraordinary new medical therapies, would not be so deeply unknowable as extraterrestrial aliens. Their psychological nature would, we might assume, be path-dependent, and the path to it would go via the evolved psychology of ordinary human beings. Nonetheless, radical enhancement as it's understood in this volume would, indeed, be radical in its immediate results and its longer-term consequences. If we greatly extend human capacities, this changes everything to a greater or lesser extent. It should not be thought of as merely adding some extras to current *Homo sapiens* and current human societies. We don't really know what we're in for with radical enhancement, and that is, to say the least, a reason to pause and consider the possibilities. Even if we're not at all technophobic, and even if we're sympathetic to the general idea of augmenting human capacities, radical enhancement is, at least in part, a step into the unknowable.

Nicholas Agar, a philosopher based at Victoria University of Wellington in New Zealand, has emphasized this point in a valuable body of work that constitutes a thorough critique of radical enhancement and its advocates. As he makes clear, we largely fail when we attempt to imagine what changes would flow from radical enhancement. It's one thing to be told that new technologies are in the offing that could make us (for example) much stronger, smarter, and longer lived. That sounds attractive, but it's another thing to imagine in detail what it might be like for us as individuals, and it's still another to think through how an entire human society – or even our entire species-wide civilization – might function with much smarter, longer-lived, and generally more capable people than any human beings in history. Agar argues, and it's worth hearing him out, that if we really understood what is involved in radical enhancement we'd reject it. We'd prefer ourselves as we are, and we'd try to improve human societies through less drastic means.

Agar has said, I think persuasively, that a philosophical interest in human enhancement, in its different degrees, is timely in what has become an age of enhancement:

> A variety of technologies promise human enhancements vastly more powerful than those provided by any pill, injection, supplement, or genetic modification available in the early years of the twenty-first century. Advances in genomics are uncovering genes that influence intelligence, longevity, and many other traits that we may want

> to enhance. [...] Would-be enhancers of humans are not restricted to genetic technologies. Some of them look to cybernetic technologies to enhance by fusing machines to human brains and bodies. (Agar 2014, 2)

In Agar's view – though I will ultimately question this – "It would be a serious mistake to just assume a scenario in which enhancements arrive in small increments with plenty of time for us to adjust between each installment" (Agar 2014, 5). As he observes in his 2010 book *Humanity's End*, "aspiring radical enhancers" want extended lifespans, enhanced intelligence, and other extreme augmentation of human capacities almost immediately: "They want these things for themselves – and if not for themselves, for their children" (Agar 2010, 5). In any event, we are already getting deeper into the game of enhancement. Even if enhancing human capacities is morally okay in principle, which I'll assume to avoid much-mapped territory (for a start, see Blackford 2014a), we face important questions. These include questions about the *timing* and *degree* of enhancement to individuals, and questions about the *pace* and *magnitude* of social change. Potentially, we're at the dawn of a great transition to a different form of civilization and a very different world. It is timely to think about what that might mean and whether we really want it.

Throughout what follows, I'll address the views of many philosophers, bioethicists, and other thinkers, so this volume is more than a lengthy response to Agar. Nonetheless, I'll pay close attention to his critique, partly because of its particular merits and partly to provide a sense of unity. There is much to disagree with, but also much that's difficult to dispute. Ultimately, Agar's prescriptions and mine may not be terribly far apart, irrespective of much wrangling over detail. Very often, the arguments on all sides depend on science fictional speculation, and while there's nothing wrong with that, it does necessitate that our attempts to imagine the future are highly uncertain. We are all struggling to some extent, and when scrutinizing others' views, including Agar's, I'll try to be fair and good-humoured.

2. Some terminology and clarifications

Before going further, we need a few definitions. In his first book about radical enhancement, *Humanity's End*, Agar introduced the concept as follows: "Radical enhancement involves improving significant human attributes and abilities to levels that greatly exceed what is currently possible for human beings" (Agar 2010, 1). He illustrates this by referring to capacities that dramatically exceed, say, the intellect of Albert Einstein, the athletic prowess of the champion sprinter Usain Bolt, or the longevity achieved by the French centenarian Jeanne Calment, who was documented as living for 122 years (from 1875 to 1997). By way of contrast, he distinguishes the idea of moderate enhancement. This might involve, for example, undergoing genetic modification to obtain prowess comparable to that of current elite athletes (Agar 2010, 17).

Though Agar is not opposed to all technologies that might enhance human capacities, much of his writing over the past decade has expressed opposition to radical enhancement as he defines it. On his account, enhancement of human capacities is a good thing, but one we could have too much of. He spells this out in the opening pages of *Truly Human Enhancement:*

> I endorse moderate enhancement – the improvement of significant attributes and abilities to levels *within or close to* what is currently possible for human beings. I reject radical enhancement – the improvement of significant attributes and abilities to levels that *greatly exceed* what is currently possible for human beings. (Agar 2014, xi; Agar's emphasis)

Agar acknowledges that there could be a grey or vague region between clear cases of radical and clear cases of merely moderate enhancement. That is, we cannot draw a sharp line between capacities that merely exceed current human possibilities and capacities that "dramatically" or "greatly" exceed them. Nonetheless, many imaginary cases undeniably fall into the region of radical enhancement. It is one thing to break the current world record for the 100-metre sprint by a fraction of a second, and a technological innovation assisting this could be regarded as producing a moderate enhancement. It would be another thing – something radical – for one of us to run with the swiftness of a cheetah.

Likewise, it is one thing to exceed historically measured levels of human intelligence. It's another to have intellectual powers that lie on the other side from us of an immense cognitive gulf. My conception of radical cognitive enhancement in the following chapters involves just such a gulf. It implies the creation of beings whose intellects dwarf those of humanity's greatest geniuses of the past and present, and who stand to us intellectually much as we stand to non-human animals such as gorillas and chimpanzees.

Interestingly, Agar does not think *genetic* enhancement – modifying human DNA to enhance physical and cognitive capacities – is the most troubling issue. In *Truly Human Enhancement*, he suggests that environmental rather than genetic enhancements should be our focus of moral investigation, since "They'll be the chief category of changes interfering with autonomous choice or fracturing the human species into haves and have-nots" (Agar 2014, 154). In this context, environmental enhancements might include a wide range of non-genetic modifications to human bodies and minds, but especially attempts to integrate our brains and nervous systems with computer technology.

In all, we can usefully bear in mind the variety of forms that enhancement might take, with more than one dimension to classifying them. In a lengthy article published in the journal *Philosophy and Public Affairs*, the American philosopher Allen Buchanan distinguishes usefully between *modes* of enhancement (such as genetic, cybernetic, and pharmaceutical) and *kinds*, or *types*, of enhancement (such as cognitive, motivational, and physical – the latter including resistance to ageing and resistance to disease) (Buchanan 2009, 351).

In *Humanity's End*, Agar identifies three specific kinds of experience that we value, and which he sees as species-relative values. That is, they are values held by *Homo sapiens* as we are now, and "are unlikely to appeal to posthumans or rational aliens" (Agar 2010, 179). They are: first, a general set of "experiences and commitments that make us who we are" (Agar 2010, 180); second, our experiences with our children; and third, elite athleticism. The argument is that radical enhancement would alienate us from each of these things. I'll address these issues, among others, in the chapters that follow.

Although some of the proposals considered in this book are, indeed, radical, my general assumption is that we could, for the foreseeable future, retain a biological form not greatly different, in its gross morphology, from what we possess currently. I assume that is, that we can look much as we do now, while attaining dramatically augmented physical and cognitive characteristics (including much greater longevity) or greatly altered (supposedly for the better) emotional or psychological tendencies.[1] Radical though this kind of enhancement might be, the idea is considerably less so than some others that enter current debates about enhancement technologies: for example, the idea that we could, through technological means, become intelligences existing in a non-biological substrate such as advanced computer hardware, while retaining our present personal identities. This raises many conceptual and scientific issues of its own, and they deserve a separate book. Thus, I will not discuss the possibility of so-called mind uploading, and I will therefore not address Agar's doubts about it.

For different reasons, I will not engage with each and every one of Agar's numerous criticisms of proposals for radical life extension. In *Humanity's End*, much emphasis is placed on the claim that radically extending our lives would worsen them in various ways, at least when judged by our current standards or supposedly human standards – though not necessarily when judged by the transformed standards of people whose lives have actually been extended. For example, Agar thinks that radical life extension would increase our fear of death. I'll come to this in due course. In *Humanity's End*, however, and especially in *Truly Human Enhancement*, Agar also argues that efforts to develop radical life extension would require unethical biomedical research. This argument is one that I plan to set aside in the following chapters, but it deserves at least a mention.

Briefly, Agar argues that no one, however young, old, healthy, or otherwise, would have sufficient motivation to enrol voluntarily as a subject in risky research aimed at extending the maximum length of human life, as opposed to curing specific, identified diseases that shorten human lives within what we currently think of as a normal span. If that is so, he suggests, research on radical life extension would predictably and unavoidably involve the coercion or exploitation of research subjects. One type of response to this might point to the current willingness of many people to try substances such as the diabetes drug metformin purely for life-extension purposes. Another response might include ingenious proposals to make it worthwhile to partici-

1 Radical physical enhancement might be a partial exception to this general assumption. We can readily imagine huge or reshaped athletes whose physiques are more extreme versions of what is currently required in elite sport. See Chapter Five below.

pate as research subjects, though this approach would soon turn into a complex debate about the nature of coercion, exploitation, and what it is for research subjects to act autonomously and to give genuine, fully informed consent. A more robust, and seemingly dangerous, response might involve biting the bullet and justifying any coercion or exploitation on utilitarian grounds.

Agar concludes that the time for concerted efforts to cure ageing, understood as a disease, lies in the relatively distant future, should a time come when a series of medical successes against the familiar diseases of ageing, such as dementia, heart disease, and diabetes, shifts our attention. Once that has happened, "people who expect to achieve a life span of only eighty years will feel that they are missing out on the century to which they are entitled", and it might then be rational to participate in clinical trials for anti-ageing therapies (Agar 2014, 136).

Generally speaking, I will treat issues as to whether we could ethically "get there" – get to the technology for creating radically enhanced people – as beyond the scope of my thinking in this book. I'm mostly concerned with what "there" might be like. I will, however, give attention in the final chapters to the question of how – and whether – we might navigate a time of transition when people who are enhanced to various degrees (or not at all) must coexist in mixed human–posthuman societies. Thus, I'm worried about a number of questions. First, would individuals find that radical life extension, or any other form of radical enhancement, is a "poisoned chalice" (Bhattacharya and Simpson 2014, 234)? Would the social uptake of radical enhancement technologies produce intolerable collective outcomes, if not in the judgement of the people living in future societies, at least by *our* standards now? Irrespective of the eventual outcome, how could we manage a transition without unacceptable conflict, or else intolerable relations of dominance–subordination, in mixed societies of the radically enhanced and the less enhanced (or unenhanced)?

3. A roadmap

There could be many impressive forms of human enhancement, and I can't discuss them all individually in a relatively short book. Consider, for example, the advantages of dramatically reduced, or totally eliminated, need for sleep, as in Nancy Kress's 1993 novel *Beggars in Spain* and its sequels. Or imagine a fabulously amped-up immune system giving immunity to almost all diseases. In what follows, however, I focus mainly on four kinds of radical enhancement, not for lack of ability to imagine others but because they are relatively obvious, receive much discussion in the existing academic literature, and feature strongly in Agar's critique. They are as follows:

- Radical physical enhancement of a kind that would be useful in athletic performance.
- Radical cognitive enhancement aimed at creating minds dramatically smarter than those of any great geniuses from the past, or from our own age.
- Radical life extension so as to overcome senescence and delay death indefinitely.
- Moral status enhancement, which involves creating beings who merit more moral consideration than existing human infants, children, and adults.

Chapter Two charts how we've become a species with a heightened sense of its own future. Prior to the Industrial Revolution, few people had any sense of a human future shaped by developments in science and technology. In fact, the ubiquitous imagery of high-technology futures is even more recent. By now, thoughtful people feel a need to take responsibility for the future, and hence for the technologies that will shape it. A problem here is that we are not very good at predicting how new technologies will be used and what their social consequences might be. Nonetheless, we see vigorous debate between supporters and opponents of various emerging or postulated technologies.

In Chapter Three, I focus on Agar's approach in his role as a leading philosophical critic of radical enhancement. (It's worth noting, however, that he does not regard himself as a "bioconservative" thinker, since he does not, in general, oppose enhancement of human capacities to what he regards as a moderate extent.) Agar relies on such concepts as species-relative values – in this case the values he believes we have in common as human beings – in order to judge the likely effects of enhancement technologies. In response, I see problems with this concept and the related concept of trying to preserve our humanity. I conclude that the real arguments should relate more to the values *actually held* by the participants now, rather than a set of supposedly cross-cultural values that could ground a universal human morality. In any event, any values shared at the species level need not necessarily include the value of remaining human. Perhaps it is in our nature as human beings to place a value on becoming, in an appropriate sense, *more than* human.

This leads to Chapter Four, which provides something of a turning point. Here, we confront the general problem of transformative choice. Agar makes much of this, frequently emphasizing that someone who undergoes radical enhancement will emerge from the experience with altered values. Even if they made an ill-informed or poorly reasoned decision, and indeed the "wrong" decision when evaluated by their values prior to the transformation, they might later be glad that they chose as they did. When evaluated using their *new* values, the decision might seem like the correct one after all. We face a conundrum here as to which values should matter in making transformative choices. Agar would recommend that we apply the species-relative values of *Homo sapiens* when the choice is about radical enhancement. I can only recommend that we (try to) use our current values, whatever they are, while acknowledging that this encounters conceptual and practical difficulties.

Chapter Five takes up specific issues relating to radical enhancement. Here, I examine the alleged threats to sport and science from radical physical enhancement and radical cognitive enhancement respectively. These threats, as presented by Agar, are exaggerated. The institutions of sport might well be changed by moderate or radical physical enhancement, but sport would undoubtedly survive, and not necessarily in a less desirable form. That said, I conclude that there is currently no good case to pursue radical physical enhancement of the kind under discussion. Accordingly, it will play no role in this book after Chapter Five. As for radical cognitive enhancement, the threat to science appears to be an illusion. The more troubling issues with radical cognitive enhancement relate to its psychological and social impact, both to be taken up in later chapters.

In Chapter Six, I examine radical life extension in detail, taking due note that it would be a transformative choice for any individual who took it up or for any society where it became common. We are not good predictors of the effects of such a radical technology, but to make this choice we must try our best to imagine the possibilities. Among other issues discussed in this chapter, I consider the argument that people who do not undergo senescence, and can therefore live indefinitely, would be saddled with a heightened fear of death. Spoiler: I think they'd fear death less than we do, but please wait for the argument.

Chapter Seven deals with a specific problem for radical life extension, one that has been known since the turn of the nineteenth century: its potentially Malthusian implications, and the issues that arise from any efforts to deal with this. The problem is Janus-faced, in that it faces toward an overpopulated world with environmental degradation and conflict over resources, but also toward a world with harsh population policies that restrict reproductive freedom.

Chapter Eight picks up a thread from the earlier chapters. The question is whether radical cognitive enhancement and/or radical life extension would be harmful or self-defeating in that they'd undermine our ongoing personal identity. Alternatively, we might survive as the same people but changed in ways that we see as undesirable prior to being enhanced. The arguments here become conceptually difficult, involving notorious unsolved problems in contemporary philosophy. That said, the cognitive and personality changes feared by opponents of radical enhancements should not necessarily be viewed negatively. As an alternative, they might represent growth and adventure.

The aim of Chapter Nine is to introduce and examine the concept of *moral status enhancement*, which means the creation of post-persons: beings with greater-than-human claims to moral consideration as a result of some kind of dramatic qualitative superiority to existing *Homo sapiens*. This contrasts with *moral enhancement*, which I introduce only briefly to define the concept and explain the difference. Moral enhancement is an important issue in itself, but moral status enhancement would seemingly involve more radical technological interventions as well as having more troubling implications.

In particular, we might worry about the social environment for ordinary human beings in a society containing post-persons. Even if post-persons are not strictly possible, perhaps because the very idea is incoherent or frustratingly mysterious, some of the radically enhanced people of the future might view themselves as post-persons or something similar. Thus, this chapter introduces fears of conflict or oppressive social relations in societies that have embraced radical enhancement.

Chapter Ten continues an investigation of social conflict and how it might be averted or minimized without suppressing radical enhancement and its possible benefits. How might radically enhanced and unenhanced (or moderately enhanced) people get along?

Despite the urgency of such questions, this final chapter includes a positive, if muted, message. I differ from most commentators on radical enhancement in that I am neither committed to preventing it, if I can, nor motivated to bring it about as quickly as possible (preferably within my own lifetime). If we're at the start of a transition to a posthuman civilization, it will not and should not take place overnight, or over a single generation. The transition will and should be slower – we might say "more stately" – than most advocates of radical enhancement seem to desire. The good news is that it need not be the end of the human story (*pace* Agar 2014, 200).

4. The future is not set

Our choices have causal efficacy, and in that sense our future is not set. The future arrives, in part, *through* us, and through our choices and actions. But when we choose and when we act, we can't reliably predict what the consequences will be. Choosing is our responsibility, but not-fully-knowing is unavoidable. We can, however, reflect seriously about possible futures and what we want from them. With luck, this might help us predict a little bit less blindly and choose a little bit more wisely. Books such as Agar's can help us think more clearly about humanity's future, and I hope to make a similar contribution.

In writing the chapters that follow, I've had to make numerous judgements and choices, such as exactly which topics ought to be covered. As a result, some have fallen by the wayside, such as mind uploading, and others have received only brief discussion, such as moral enhancement. My best guess is that the former will not be feasible and the latter will not greatly impact on humanity's future even if it's developed and widely used. I might be wrong on both counts – I'm a seriously imperfect predictor like everyone else – but I suspect that attempts to develop these technologies will not alter the overall picture.

In a previous book, *Humanity Enhanced* (Blackford 2014a), I defended many uses of genetic technology, including certain uses for what Agar would regard as moderate enhancement. This time, the stakes are higher, since radical enhancement is, after all, *radical.* It shouldn't be surprising then, that I strike a more cautious note on this occasion. Nonetheless, I do not reject the idea of radical enhancement. Rather, I have a different theme: the path to radical enhancement must go via moderate enhancement. That won't please all my friends, some of whom want radical enhancement – especially radical life extension – as quickly as possible, preferably yesterday, or last year, and certainly *right now.*

Chapter Two will provide a sketch of how human beings discovered the future. We now see passionate advocates of rival futures for our planet and our species, among them transhumanists and their critics. Who are these people, where did they come from, and who made a world that has such people in it?

Chapter Two : The Very Idea of the Future

1. Beings in time

Writing in the seventeenth century, the English philosopher John Locke conceived of personhood as the combination of three specific properties: the capacity for reason; awareness of oneself as an individual being; and a sense of oneself as existing in time, with a past and a future. Unlike most or all other living things, human beings typically possess these properties, and this enables us to develop complex personal relationships and equally complex preferences about future events. This also entails that we are vulnerable in certain ways, above and beyond the vulnerability of other creatures that can experience pain and other sensations, but do not have ambitions, plans, hopes, and fears for their own futures. I'll return to this theme in Chapter Nine.

For now, I should acknowledge that some non-human animals might have these Lockean properties at least to a certain extent. Conversely, even human beings who possess the Lockean properties might not actually have any complex preferences or plans. Nonetheless, typical human beings – children beyond a certain level of development as well as adults – usually have a highly developed sense of themselves as creatures who exist in time, with a future as well as a past.

And yet, humans have not always had a comparable sense of themselves as a *species* existing in time with a future that might be very different from the past and the present. To place this in perspective, our species, *Homo sapiens*, is 200,000 to 300,000 years old, and earlier human species from which we descended go back much further, millions of years into the past. The rise of agriculture dates back about 12,000 years, and something recognizable as civilization, with large cities, writing, and bureaucratic social organization, emerged in the Middle East and other locations roughly 5,000 years before the present day.

What we now call European modernity is historically recent. If we could travel back to Europe in, say, 1500, or even 1600, CE, we'd find societies in which there was little sense of ongoing social change. Changes in work methods and transport – and even in methods of social and political organization – happened too slowly to be transformative within a single lifetime. There had always been large-scale changes from specific events such as wars and conquests, plagues and famines, and various other kinds of human-caused or natural disasters. Some religious and mythological systems described grander cycles of time than years, seasons, and generations. Some pre-modern societies believed in a lost golden age from which they'd degenerated, and Christian writings prophesied an eventual end of worldly things to be brought about by the intervention of God. But none of this resembles our contemporary idea of the future, in which human societies are continually transformed by new technologies and advances in scientific knowledge.

2. Modernity and the human future

As the scientist and social commentator J. D. Bernal observed in the 1920s, human beings normally take accidental features of their own societies to be axiomatic features of the universe, likely to continue until supernaturally interrupted: "Until the last few centuries this inability to see the future except as a continuation of the present prevented any but mystical anticipations of it" (Bernal 1970 [1929], 12). Prior to European modernity and the rise of science, humans might have expected supernatural events in the future, such as the second coming of the Messiah, but they did not imagine events such as the invention of the steam engine, the spread of the railways, electricity, the telegraph, motor cars, and aviation.

That said, the sixteenth century in Europe was an extraordinarily volatile period. It immediately followed the invention and spread of the printing press and the European discovery of the New World. The century began with festering religious discontent that led to the Protestant Reformation, and then to Europe's great wars of religion that extended, in one form or another, deep into the seventeenth century. The sixteenth century also saw the beginnings of modern science, including the radically innovative astronomical work of Copernicus.

The seventeenth century brought scientific investigation to a new level. Geniuses of the Scientific Revolution employed increasingly sophisticated techniques for systematic and quantitatively precise investigation of the natural world. Sciences such as physics and astronomy took advantage of advances in mathematics, the design of controlled experiments, precisely engineered equipment, and new instruments of observation such as Galileo's telescope. Much of this added greatly to the power of hypothetico-deductive reasoning – the process of formulating and testing conjectures – to draw reliable conclusions about mechanisms that could not be observed directly. In the early decades of the seventeenth century, there was a ferment of new methods, together with a sense of cooperation, rivalry, and excitement. This era was soon followed by a series of impressive technological innovations such as precision clocks (which had far-reaching benefits for navigation and trade), the steam engine, and the power loom. It appeared that the rise of science had laid a foundation for shaping and controlling nature, which prompted a further thought:

> Without much difficulty, the thought followed that, if nature can be induced to do our bidding, why should not human society, of which we are after all the creators, be made to do the same? (Johnston 2011, 169)

As the sciences developed, their practitioners were able to study a great range of natural phenomena that had previously resisted human efforts. The rise, consolidation, and extension of science, perhaps more than anything else, led European thinkers of the eighteenth-century Age of Enlightenment to imagine future states of society with superior knowledge and wisdom. The upheavals of the French and American Revolutions inspired even more conjectures and schemes involving future societies. The new way of thinking is reflected in the work of many early nineteenth-century thinkers, such as Henri de Saint-Simon. For David Johnston, however, Karl Marx's writings were its apotheosis:

> Writing in the middle years of the century, Marx argued that human history up to that time had been made with no human awareness of the broad, systemic consequences of human actions. As a result, the major forms and patterns of social cooperation had over the course of history taken on a succession of highly consequential shapes that no human being had ever consciously designed. (Johnston 2011, 170)

With few exceptions, however, Enlightenment ideas of intellectual, moral, and social advances showed little of our emphasis today on new technology when we try to imagine the future. Even in 1800, and for some time thereafter, there was little emphasis on *technological* transformations of society, but our current idea of the future developed gradually and gathered force, responding to the Industrial Revolution, which commenced during the second half of the eighteenth century, at first in Britain, but then in other European societies and their colonies. As it continued and renewed itself, changes in the ways things were done happened on a large scale and at a pace that could not be ignored by well-informed citizens. Countless thousands of years after its emergence in Africa, our species, *Homo sapiens*, gave birth to a new kind of change: continual, visible change, driven and shaped by science and technology. We might reasonably say that the nineteenth century was when humanity discovered the future.

3. A technological future

It is fascinating to observe the evolution, throughout the nineteenth century, of what would now be called science fiction narratives. For example, Mary Shelley's *The Last Man* (1826), unlike her earlier *Frankenstein; or, The Modern Prometheus* (1818), does not depict changes of a dramatically new kind, produced by technology. *The Last Man* is set at the end of the twenty-first century, long after its author's own time, but it portrays the economic and technological base of society as essentially unchanged. No transformations have been produced through technology: for example, there is nothing like the prospect of a new breed of artificial humans waging war upon ordinary humanity, such as Victor Frankenstein fears in Shelley's earlier novel. Instead, humanity is destroyed by a global plague. This is a natural, rather than supernatural, end to human history, but it is also a grand-scale example of the epidemics that had swept across Europe in previous centuries and even in Shelley's own lifetime.

In 1827, *The Mummy! A Tale of the Twenty-Second Century* was originally published anonymously in London. At the time she wrote and published it, the novel's author was Jane Wells Webb, known after her marriage in 1831 as Jane Loudon. *The Mummy!* is a melodramatic – and often comical – novel set in 2126, three hundred years after the time of its composition. Like *Frankenstein*, it is something of a transitional work between the gothic style and science fiction as we know it today.

The Mummy! depicts a future Europe somewhat changed by technology: mainly the existence of electricity, and flight in a variety of balloons and other airborne vehicles. As we learn as readers, balloons can quickly traverse vast distances at a constant latitude, since they can reach great heights where, in accordance with the book's weird science, they are able to become stationary while the earth continues to rotate beneath them. In general, however, technology has made little mark on social arrangements – little more than in Shelley's *The Last Man*. Central to *The Mummy!*'s narrative is the revival of an ancient Egyptian Pharaoh, Cheops, apparently by means of a galvanic battery (here, *Frankenstein* was undoubtedly an influence). Eventually, however, it transpires that Cheops was unleashed on the world by supernatural design, to assist the innocent and lure the wicked to destruction. It seems that the real cause of his awakening was the power of God rather than that of human technology.

By contrast, the science fiction of Jules Verne, later in the nineteenth century, is informed by a trust in science and progress that had increasingly come to predominate – although even here we can discern an element of ambiguity. For example, Robur the Conqueror, the eponymous protagonist of *Robur the Conqueror* (1886; also known in English as *The Clipper of the Clouds*) and *Master of the World* (1904), dies in a last act of hubris, fighting, quite unnecessarily, against the might of a hurricane in his superscience air/sea/land vehicle the *Terror*. Likewise, the *Nautilus* of Captain Nemo is wondrous and alluring, but also a menace to shipping. At least in his first outing, in *Twenty Thousand Leagues Under the Sea* (1870), Nemo is not so much a hero as a charismatic supervillain.[2]

By the 1880s and 1890s, scientific romance (the early form of modern science fiction) was becoming a genuine fiction of the future, to borrow a phrase from the literary scholar Robert Scholes (1975). Authors such as H. G. Wells had reached a new understanding of history and technological

2 Nemo is a more benevolent figure in *The Mysterious Island* (1874–1875), after surviving the earlier book's climactic events.

change. In 1939, during the immediate build-up to World War II, Wells reflected on what he perceived as a striking modification of humanity's understanding of time since the generation of his grandparents. As he saw things, human beings in the late nineteenth century had possessed an "extremely shallow" sense of the scale of time, and hence "scarcely any historical perspective at all" (Wells 1939, 22). They had imagined the beginning of time to be only a few thousand years before them, had understood human life as essentially unchanging, though perhaps spreading, and altering in particulars, and had been unaware of human-caused environmental degradation that took place too slowly to be apparent within a single lifetime. By contrast, Wells and his readers had a new perspective:

> We know that the everlasting hills are not everlasting, that all our working conceptions of behaviour and destiny are provisional and that human nature and everything about it is being carried along upon an irreversible process of change. (Wells 1939, 24)

For Wells and his peers, it was possible to contemplate the very ancient beginnings of the human species, to know something of the early development of *Homo sapiens*, and to trace the emergence of hunter-gatherer society, agriculture, and the rise of urban civilizations and their accompanying languages, religions, and systems of law. It was apparent that humanity had, "for the last twenty or twenty-five thousand years", undergone a continuously accelerating process of change, unprecedented among other forms of life (Wells 1939, 25). For Wells, this suggested the further thought that there was no guarantee of humanity's survival. From the new perspective, we are subject to the same natural processes that forced the extinction and evolution of other species, and our future, if we do survive, will not simply resemble our past. Hence, it made sense for Wells to produce a book with a title such as *The Fate of Homo Sapiens* – whereas such a title would once have been meaningless outside the context of supernatural prophecy.

As Wells knew, the rapid changes of the nineteenth century implied the likelihood of rapid – perhaps *more* rapid – changes to come. He thought, at least for most of his career, that it was possible to consider and imagine the future of humanity with some prospect of making successful predictions. He discussed this topic in a celebrated lecture, "The Discovery of the Future", that he delivered to the Royal Institution of London in January 1902. This helped to establish his reputation, and the lecture was published as a small book (Wells 1902) not long after. For Wells, the present had arisen from the past through the deterministic operation of scientific laws, and the future would be constrained in the same way. However, he pointed out an asymmetry between the past and the future, or at least in how we perceive them. That is, we can be certain about many events that happened to us personally in the past, and which we remember clearly, whereas we do not know what lies in store for us, as individuals, in the future. We have no future-oriented equivalent of personal memory.

But, he added, things are different when it comes to future events involving large populations. Wells acknowledged the impact of exceptional individuals, but was strongly inclined – not unlike Marx in this respect – to think about larger forces operating in history. For example, if Julius Caesar or Napoleon had never been born, someone else would have played a similar role in the history of Europe and produced the same broad outcome. Accordingly, he concluded that we have evidence available to us in the present that can help us to reconstruct the past and to predict how humanity's future will unfold on a large scale.

Wells saw clearly and thought deeply, but is the future really discoverable in the sense he claimed? Now that we have a better understanding of technological and social change – much better, say, than that of Wells's grandparents – we crave knowledge of how things might turn out, and yet the future is still largely unknown. Wells had some impressive predictive successes, but even these were limited. He foresaw the use of armoured military vehicles in his short story "The Land Ironclads" (1903), the importance of aviation for future warfare in several works including *The War in the Air* (1908), and, in *The World Set Free* (1914), the development of massively destructive atomic bombs. But none these developments turned out quite as he imagined them. It is one thing to have a scientifically informed sense of humanity's future; it's another to predict events with much detail or accuracy.

Earlier, I suggested that the nineteenth century was when humanity discovered the future. All the same, there was some way to go. Conceptions of a technological future spread and developed during the century that followed. In a book-length study of science fictional anticipations of twentieth-century super-weapons, such as atomic bombs and rocket-propelled missiles, the astrophysicist and scholar Andrew May examines the Cold War's effect on the popular imagination. He comments that speculations about the future earlier in the twentieth century were still likely to focus on political and military conflict and on social change, and not so much on science and technology. It seems that the perspectives of Bernal, Wells, and relatively few others needed time before they soaked in and became commonplace.

While science fiction writers of the 1930s and 1940s "dealt with such topics [as space travel and nuclear weapons] on a routine basis" (May 2018, 195), the wider public became aware of these topics somewhat later, when terrifyingly advanced military technology began to reshape the global order. As a result, the Cold War, with its competition between great powers for technological and military superiority, consolidated humanity's process of discovering the future. By the 1960s and 1970s, what had once seemed like scientific marvels were commonplace and making daily headlines. When *we* think about the future we are likely to take it for granted that it "will be defined by its technology" (May 2018, 195), but that's a recent development in human history.

In the twenty-first century, we see many speculations and anxieties about humanity's future. Some visions of the future involve doom or despair, while others are radically optimistic. However we imagine the future as individuals, we've become a self-conscious species, aware that our collective actions have large consequences, for good or ill, though we can't be sure in advance what they might be.

4. Science meets the shadow of mortality

Prior to the rise of modern science, there was already a long history of myth, magic, and speculation about greater-than-human powers, and especially about the possibility of immortality or extreme longevity. In his classic study of ideas about prolonging human life, Gerald A. Gruman makes a standard distinction between life expectancy and the maximum human lifespan – in the sense of the length of life of the longest-lived individuals. He assesses the latter as about 110 years, which remains a good estimate nearly 60 years later, though in the interim Jeanne Calment set a new record of slightly over

120 years.[3] Calment notwithstanding, Gruman correctly observes: "Unlike life expectancy, the span of life does not seem to have increased noticeably during the course of history" (Gruman 1977 [1966], 7).

Gruman uses the term "prolongevity" to refer to "the significant extension of human life by human action" and introduces the awkward term "prolongevitism" for the idea that prolongevity is possible and desirable (Gruman 1977 [1966], 6). He explains that the focus of his book is on natural prolongevitism, on efforts to extend the span of life by natural rather than supernatural means, although he also indicates that he will not draw a clear line between natural and supernatural, given the nature of most efforts before about 1800 (Gruman 1977 [1966], 9). He examines attitudes to prolongation of life in Western and Near Eastern traditions, going back as far as the *Epic of Gilgamesh*, and also in Eastern thought, especially within the traditions of Taoism. Throughout, he draws a contrast between prolongevitism and what he refers to as apologist material, in the sense of writings that bid us to accept mortality. Broadly speaking, the mythology and philosophy of classical antiquity was apologist in its tenor, as were the Hebrew Bible, the New Testament, the writings of the Church Fathers (notably Augustine and Aquinas), and those of Avicenna in the eleventh century.

It is fair to conclude that Western and Islamic myth, religion, and philosophy have generally been apologist in Gruman's sense. They've portrayed the shadow of mortality as somehow a good thing. Yet there is also a long countertradition. Gruman discusses Western legends and folklore that seem to convey more positive attitudes to very long life. This material includes antediluvian, hyperborean, and fountain-of-youth themes.[4] Thus, we can observe a contrast between the West's more socially prestigious writings – which have usually been apologist – and the yearnings for immortality or longer life given expression in legend and folk stories:

3 Calment's record currently seems genuine, although some researchers have claimed that her daughter took over her identity in the 1930s.

4 That is, the ideas that longevity was much greater in the distant past, long life can be found in distant places (beyond the north wind), and various substances and objects have magical or natural powers to extend life.

> In this chapter we have seen that, despite the rationalizing of the apologists, there existed a great profusion of legendary material expressing the basic human wish for prolonged life on this earth. In folklore, with its free expression of wishes, the basic aspirations and dreams of the unlettered populace were able to come to the fore. [...] It is as though apologism controlled the Super-ego, while meliorism and prolongevity subsisted in the half-conscious twilight of the Id. (Gruman 1977 [1966], 28)

Likewise, Gruman finds a division of thought within Taoism. Apologist sentiments about death feature in classical Taoist texts, but Taoism also includes a prolongevitist tradition. Indeed, Gruman portrays the ambitions of pre-scientific or proto-scientific prolongevitists in the West, such as Roger Bacon in the thirteenth century and Luigi Cornaro in sixteenth-century Venice, as relatively modest compared with Taoist adepts who sought true immortality.

Roger Bacon was the great figure of his time in popularizing the usefulness of alchemy for prolonging life. However, he sought extension of life by "only" a century beyond the common lifespan of the time – which suggests a life of not much more than 150 years. There was a possible implication that this might become longer in later generations, but there was an outer limit as a result of the Fall recounted in the Book of Genesis (Gruman 1977 [1966], 64). Cornaro and other "hygienist" exponents of the prolongation of life advised simplicity and temperance in diet and general lifestyle, which might be good advice, but was never going to extend the human lifespan beyond its observed limits.[5]

It is likely that all human societies during recorded history have included apologist positions on death, but they have also revealed prolongevitist yearnings, whether in proto-science and science or in the ambiguities of folklore.

As Gruman tells the story, scientific approaches began to predominate in the West as early as the seventeenth century. The idea of secular progress gradually ousted old approaches to the extension of human longevity. In the writings of René Descartes, at the dawn of the Scientific Revolution, we can find references to the possibility of curing the debility of age or retarding its

5 As it happened, Cornaro lived to a remarkably ripe old age: he was probably born in 1467, and he died in 1566.

progress. Descartes chafed at the brevity of life, and particularly at the frustrations in making intellectual progress within a human lifetime (Gruman 1977 [1966], 80–83). Unfortunately, he died in 1650 at the age of 53. This was not so bad for his era, but it fell far short of his hopes and was untimely for one of Europe's true geniuses of science and philosophy.

Likewise, Francis Bacon had a profound interest in the scientific extension of human life. His own death, in 1626, by pneumonia, may have been triggered by an experiment in using snow to preserve meat. Whether he had life extension in mind is not clear, and the story's reliability is doubtful. What we can say more confidently is that the early progress of science in the seventeenth century depended on a number of intellectual tendencies of the time, including a revived interest in ancient atomist theories and a conviction that human welfare could be improved by technological means. In the writings of Descartes, Francis Bacon, and other seventeenth-century scientists such as Robert Boyle, this turned quickly and naturally to thoughts of using science to prolong human life.

These legendary figures of the Scientific Revolution looked back, as David Boyd Haycock (2008) ably explains, to the immense spans of life supposedly enjoyed by Adam, Methuselah, and other patriarchs from the Book of Genesis, but they also looked forward to wonderful discoveries. By the end of the century, their various efforts to extend life had come to nothing, and the topic seems to have been of less interest through most of the eighteenth century. However, Gruman adds the useful point that Descartes understood the human body as a machine, which stimulated advances in biology and medicine (Gruman 1977 [1966], 79–80). This understanding of bodies as machines, and therefore open to beneficial intervention through technology, became more prevalent during the Enlightenment, and its implications still inspire advocates of both moderate and radical human enhancement.

5. Condorcet's vision of human destiny

Despite the efforts of seventeenth-century scientists, familiar ideas of secular progress became dominant only as science and technology developed an impressive track record during the following centuries. In the late eighteenth and early nineteenth centuries, the period that included the French Revolution, there was a strong sense of optimism and openness to change among the European intelligentsia. As John Avery explains, Enlightenment thinkers imagined an inevitable process of improvement in the human condition:

> The *philosophes* of the Enlightenment visualized history as a long progression towards the discovery of printing, and of the scientific method. Once discovered, these techniques could never be lost; and they would lead inevitably (it was believed) to the material and moral improvement of society. (Avery 1997, xiv)

Writing in the 1790s, the French philosopher, mathematician, and statesman Antoine-Nicolas de Condorcet examined humanity's social, technological, and intellectual history from the earliest known times. In the final chapter of his *Sketch for a Historical Picture of the Progress of the Human Mind*, Condorcet considered the prospects for forecasting "the future destiny of man" (Condorcet 1979 [1795], 73). Like H. G. Wells over a century later, he suggested that it should, in principle, be possible to predict the broad outlines of future history based on an analysis of past experience. Condorcet expected increasing equality between and within nations – including the equality of men and women – predicted the retreat of European colonialism, and advocated the development of science and the universal extension of education. On this basis, he foresaw immense progress in the sciences, technology, industrial productivity, the fine arts, moral and political ideas, and human welfare as a whole. Not least, he predicted the eventual abolition of war.

For Condorcet, the ultimate question related to humanity's perfectibility, and he considered whether this might be constrained by the carrying capacity of our planet – in the sense of an eventual limit on food production and hence a need to limit population. In that respect, he predicted an end to "the absurd prejudices of superstition" that he saw as corrupting and degrading the moral code of his time (Condorcet 1979 [1795], 189). That is, he imagined an eventual policy of controlling population, on the basis that what is owed to the unborn is not that they come into existence but that those who do come into existence have prospects for long life and happiness. These comments, among others, provoked Thomas Malthus to present a far more pessimistic analysis of population growth in *An Essay on the Principle of Population*, published in its first edition a few years later in 1798.

The final pages of Condorcet's book discuss social and medical improvements that will benefit human health, physical strength and dexterity, perceptual, intellectual, and moral abilities, and the span of life. Here, the author imagines ongoing increases to human longevity with no limit in sight, perhaps to the extent that individual deaths become extraordinary occurrences. With this radically optimistic vision of the future, he offers comfort to all who work for "the progress of reason and the defence of liberty" that their efforts are "part of the eternal chain of human destiny" (Condorcet 1979 [1795], 201) – a thought that evidently comforted Condorcet himself as he worked on the *Sketch for a Historical Picture of the Progress of the Human Mind* while in hiding from the revolutionary authorities during the bleakest period of the French Revolution.

Condorcet died in prison (at only the age of 50) in March 1794, shortly after being arrested in the village of Clamart outside the gates of Paris. A mere four months later, the revolutionary leader Maximilien de Robespierre was arrested and executed, bringing the Reign of Terror to an end. Published posthumously by the French government in 1795, Condorcet's *Sketch for a Historical Picture of the Progress of the Human Mind* became a leading and influential statement of liberal thought, and especially the ideas of progress and a magnificent human destiny. As Avery states, the *Sketch* "was destined to establish the form in which the eighteenth-century idea of progress was incorporated into Western thought" (Avery 1997, 12). For Condorcet, it was intended as merely the prelude to a much longer work that would have contained more in the way of argument and detail, but posterity was robbed of this by his persecution and early death.

Condorcet's *Sketch* defined the idea of progress in French thought, but in England that role was played by William Godwin's massive *Enquiry Concerning Political Justice* (1793), which "caught the imagination of the English-speaking world" (Avery 1997, 13). Godwin's *Enquiry* subjects the moral and political understandings of his time and place to exhaustive and radical re-examination, discarding much of the accumulated wisdom – if such it was – of the age, and advancing new principles. In an appendix to the volume's penultimate chapter, Godwin speculates about the indefinite prolongation of human life and vigour. He imagines this taking place not through physical interventions, using new technologies, but via intellectual improvement and the eventual extension of conscious control over our impulses, and even over hitherto involuntary bodily functions such as the circulation of the blood.

During the nineteenth century, life expectancy increased due to developments in sanitation, vaccines, antisepsis, and anaesthesia, among other advances, while pre-scientific and proto-scientific theories of life prolongation were undermined by the debunking of popular claims about super-centenarians who had lived for 150 years or more. For prolongevitists, all hope now lay in science:

> It appeared that God and Nature had not been so generous in allotting years to man, and if he desired a greater longevity, he would have to achieve some significant breakthrough in science and medicine. (Gruman 1977 [1966], 74)

At this stage, we should also note some pessimistic strains through the nineteenth century. While Condorcet's vision was influential, so was that of Malthus. Thus, the author and futurist Jennifer M. Gidley discusses the Malthusian analysis as triggering pessimism and anxiety about the human future, leading to bleak narratives such as Shelley's *The Last Man* (Gidley 2017, 37–38). Malthus also helped inspire the Darwinian theory of natural selection, and this, in turn, was interpreted pessimistically by some authors in the late nineteenth century who feared – or even believed they could see – a degeneration of the human species. This despairing thought inspired Francis Galton's ideas of eugenics, and it achieved its strongest statement in Max Nordau's *Degeneration* (1892), published almost a century after the great optimistic works of Godwin and Condorcet, but with the opposite message.

Haycock observes: "Civilization, it appeared to some critics, was turning backwards before their eyes. The nineteenth century ended on a downbeat note" (Haycock 2008, 157). This last claim is an exaggeration, in that plenty of optimism remained at the end of the nineteenth century – much of it to be dashed by the horrors of trench warfare in World War I and by the later atrocities of aerial bombardment, nuclear warfare, and the Holocaust during World War II. One way or another, however, the full-scale vision and ethic of human destiny urged by Condorcet were somewhat obscured by nineteenth-century developments.

6. Human enhancement and its enemies

The complexities of twentieth-century intellectual history, even confined to the English-speaking liberal democracies, defy useful simplification. It must suffice to say that various events forced the political classes and the general public to take scientific and technological progress more seriously, and to appreciate its potential to improve the human condition. Condorcet's optimism about human destiny can be found in much twentieth-century science fiction and in non-fiction writings by J. D. Bernal, J. B. S. Haldane, and others in the century's early decades. Much more recently, a similar vision of human destiny has inspired the transhumanist movement, which has internal variants and tensions but generally favours human enhancement, including radical life extension.

The terms "transhumanism" and "posthumanism" are often used interchangeably, but this causes confusion and is worth resisting in the interests of clarity in public dicussion. To simplify, but not too far, posthumanism is a philosophical movement that aims to go beyond humanism (however, exactly, this contested concept is understood) by developing a post-anthropocentric understanding of humanity's place in the natural order. With this goes one or another post-anthropocentric approach to ethics. Though some transhumanists do, indeed, embrace post-anthropocentric metaphysical, moral, and political ideas, these are not transhumanism's essence. Rather, transhumanism is defined by its perceived moral imperative to enhance human capacities, including, or especially, by direct technological interventions.[6] One cause of confusion is that transhumanists often refer to "posthumans" or to a "posthuman" state of being, but this is not a rejection of *humanism* (or an attempt to get beyond it) so much as a wish to create, or become, beings who are free of many human limitations.[7]

---

6 For a manageable, yet comprehensive, introduction to transhumanism, see Roberto Manzocco's *Transhumanism – Engineering the Human Condition* (Manzocco 2019) and the writings collected by Max More and Natasha Vita-More in their authoritative *The Transhumanist Reader* (More and Vita-More 2013).

7 The "trans" in "transhumanism" refers primarily to a transition from normal human capacities to a more capable posthuman state of being.

The twentieth century also saw contrary trends toward questioning scientific meliorism and technoscientific progress. Some questioning came from traditional Christian thinkers, such as the British novelist, scholar, and theologian C. S. Lewis, writing in the 1930s and 1940s. Lewis frequently targeted such figures as Wells, Haldane, and the visionary science fiction author Olaf Stapledon. He could be seen as one equivalent to today's bioconservative figures, opposing what he saw as dehumanizing trends in many scientifically informed speculations about the future. But Lewis was hardly alone. Many twentieth-century events, not least the industrialized killing techniques of the Holocaust and the nuclear bombing of two Japanese cities, encouraged pessimism about science, technology, and humanity's future. The use of technology to commit atrocities and war crimes, combined with Malthusian issues and justified environmental fears, generated widespread anxiety and even rejection of much technology.

And yet, as the Australian author and futurist Damien Broderick expresses the point in *The Spike*, we now live in a world of "runaway progress" (Broderick 2002, 22). Broderick identifies numerous trends that make radically meliorist or transhumanist hopes seem reasonable. He discusses the spectacular increases in the speed of travel and transport, ongoing and enormous growth in computer power, the prospect of human-level Artificial Intelligence (with computer superintelligence not far behind), the possibilities – seemingly without limit – of molecular nanotechnology, and many other, more speculative, ideas. Among the latter are the prospects of radical life extension, mind uploading (and living thereafter in rich virtual-reality environments), and – who knows? – perhaps making contact with unimaginably advanced extraterrestrial civilizations.

Though we cannot predict the future trajectory of any particular set of technologies, Broderick's idea of runaway progress seems apt. We should, admittedly, keep in mind that the exponentially increasing power of any particular technology might eventually become a less exciting S-shaped trend curve, with its steep slope flattening out. But even then, new technologies often arise, and technological capability in a particular domain might continue to increase rapidly, as did transportation speed for many decades after the first steps in aviation. When it comes to our ability to augment ourselves using technologies that enter our bodies and boost their potential or actual capabilities, it seems likely that we are not yet near the top of a flattening curve. Our ability in this broad domain is likely to grow in the years, decades, and centuries ahead:

> We can expect that advanced technology will provide physical rejuvenation and perhaps immortality (except for death by accident or choice), increased intelligence and emotional depth and range, machine-enhanced memory, and the power to realize in virtual reality anything we are able to imagine. (Broderick 2002, 237)

Such prospects attract enthusiastic supporters, but they also provoke a range of anxieties. The required technologies might be difficult to achieve, and some might lie beyond our grasp. Some technologies might fall into the wrong hands and be used for destructive purposes. Some might cause other problems.

In a recent book that supports radical life extension, John K. Davis offers a useful but slightly distorted picture of the recent crop of bioconservative figures and their concerns. He states: "Concerns about enhancement first surfaced in public discourse in the late 1990s and early 2000s among a group of bioethics thinkers sometimes known as the bioconservatives" (Davis 2018, 219). As examples, he lists Leon R. Kass, who was an especially prominent figure in the debates about human cloning in the late 1990s and the early years of the new century, the political scientist Francis Fukuyama, the prominent Harvard philosopher Michael J. Sandel, and the great German philosopher and social thinker Jürgen Habermas. Davis notes correctly that not all of these are politically conservative, but they and others appear to share an agenda of preserving human nature, including the human genome, largely in its natural state prior to any direct technological intervention.

Though Davis views these bioconservative figures as reacting to the views of "a group known as the 'transhumanists' or 'posthumanists'" (Davis 2018, 220), transhumanism was still relatively obscure in the late 1990s. It would be more accurate to see the bioconservatives as reacting to the astonishing announcement, in early 1997, of Dolly the sheep, the famous product of a team of researchers at the Roslin Institute in Scotland. Dolly was the first mammal to be cloned from a somatic cell extracted from the body of an existing adult mammal.[8] Although this was a dramatic breakthrough in reproductive science, with important implications for agriculture in particular, it was immediately apparent that the same technique could, in principle, be used to create genetic twins of adult human beings.

The theoretical prospect of this sort of high-tech reproduction triggered a political backlash. It led to numerous laws against human cloning and attracted much attention to bioethical issues more generally. These dominated news coverage in many countries until they were eventually overshadowed as a topic for urgent public discussion by the jihadist attacks on the Pentagon and New York's World Trade Center that took place in September 2001.

The individuals mentioned by Davis as bioconservatives – Kass, Fukuyama, Sandel, and Habermas – were not, with the partial exception of Kass, established bioethicists. Rather, they were academics and intellectuals who took a keen interest in the issues from outside the bioethics establishment, with which they entered into a tense relationship, receiving considerable criticism (as well as some support) in the leading Anglophone bioethics journals. When US President George W. Bush established the President's Council on Bioethics in late 2001, with Kass as chair and Sandel and Fukuyama as members, this was, in effect, a rebuff to mainstream bioethicists in the United States. It showed a president seeking expert advice on bioethical issues, and especially issues relating to enhancement technologies, from outside the usual professional circle. Predictably, the President's Council issued reports that opposed innovations such as human reproductive cloning and attempts at even moderate human enhancement.

Support for the leading bioconservatives came from many directions including the environmentalist Bill McKibben, many thinkers on the Christian Right – permanently opposed to much in the way of secular modernity and what they saw as its hubris – less explicitly religious thinkers focused on ideas of human dignity and its violation, individuals and organizations concerned about risks from new technologies, and some feminists and other left-wing or progressive activists. Among academic philosophers and bioethicists, there was more nuance, and indeed we can see a considerable split in opin-

8 A somatic cell is simply any cell other than a sperm cell or an ovum. Somatic cells are spoken of in contrast to gametes or germ cells.

ion between those who favoured some forms of moderate, or even radical, enhancement of human capacities and others who were more hostile.

As a prominent academic philosopher with a deep interest in bioethical issues, Nicholas Agar has developed a carefully qualified defence of what can be called technologies of genetic choice, focusing on reproductive cloning, pre-implantation genetic diagnosis (or PGD), and genetic engineering of human embryos for the purpose of therapy or enhancement.[9] However, he takes a different view when discussing extreme changes to human attributes such as cognitive ability, athletic performance, and longevity. In short, he does not object to human *enhancement*, but specifically to *radical* human enhancement:

> I conjecture that the most dramatic means of enhancing our cognitive powers could in fact kill us; that the radical extension of our life spans could eliminate experiences of great value from our lives; and that a situation in which some humans are radically enhanced while others are not could lead to a tyranny of posthumans over humans. (Agar 2010, 11)

7. A radically enhanced future?

Over the past few centuries we humans have become increasingly conscious of ourselves as a species with a future very different from the past. During that time, a succession of revolutions in science and technology greatly altered the knowledge base, and with it they transformed previous understandings of our place within the natural world. They likewise transformed the methods of industry and production, the arrangements for social organization, and, perhaps less obviously, the existing moral norms and varieties of cultural expression across the full range of human societies.

This newfound sense of a secular, technology-driven human future has created a pressure – moral and emotional – to try to *control* humanity's destiny. This has provided a theme for many books, films, articles, debates, and political ideologies over the past two centuries or more. In some respects, the

9 This qualified defence of technologies of genetic choice is the main theme of Agar's book *Liberal Eugenics* (2004).

pressure is beneficial. We are now far more aware of unexpected and undesirable outcomes from technological and other changes, which means that we now give more thought to the worst outcomes of human activities, such as catastrophic climate change, in an attempt to avoid or correct them. Conversely, we are far more aware of the contingency of the *mores*, methods of production, and forms of social organization that we've inherited. There's a wider sense that all of these could be changed if better alternatives could be imagined and developed along practical lines.

We cannot, however, predict or control much in the way of detail, even if we feel we should. Our Pleistocene ancestors were in no position to predict our current work arrangements and methods of production, our ideas of fun and leisure, or our modern ideologies and moral systems. In all those respects, our more recent ancestors in the late middle ages and early modernity were not much better off. *We* may not be much better off when we try to predict even the medium-term future, over the next two or three centuries, or to imagine the thoughts and practices of the generations who'll come after us, growing up and being socialized in environments different from our own. Needless to say, it is even more difficult to predict and imagine the feelings, thoughts, and actions of radically enhanced future people.

This might itself be a reason for caution about radical enhancement, though it's also a reason for some scepticism about alarming conjectures such as Agar's. In Chapter Three, I start to examine the case against radical enhancement, beginning with Agar's conception of humanity as a species, his idea of species-relative values, and how these relate to his urging that we try to retain what he understands as our humanity and our "human" ways of life.

Chapter Three : Agar on Radical Enhancement – Humanity and its Limits

1. Two little words – "threatens" and "deserve"

Nicholas Agar is a moderate, thoughtful, and persuasive opponent of radical enhancement, so his case against it provides a useful reference point for discussion. In *Humanity's End*, he distinguishes radical enhancement from what he calls moderate enhancement, which might include tampering with a child's genetic potential to enable them to become as smart as the greatest scientists or as athletic as the greatest boxers or tennis players. Here, there is no doubt that the child is still human. By contrast, says Agar, "Radical enhancement threatens to turn us into fundamentally different kinds of beings, so different that we will no longer deserve to be called human" (Agar 2010, 2).

Let's stop to consider two of the words in this short passage: the words "threatens" and "deserve". Without more explanation, Agar's use of these words is unearned and prejudicial, since, first of all, it is not obvious that the prospect of becoming beings of a different kind should be viewed as a threat rather than, say, a promise. Likewise, the suggestion that we will no longer deserve to be called "human" suggests that we will deserve to be called something that sounds less like a term of approbation ("sub-human", perhaps). But this is counterintuitive. Prior to considering Agar's arguments, we might think that becoming much smarter, more atheletic, and longer lived would be beneficial rather than harmful. If so, it is hardly something that we can be *threatened* with. Agar's way of making his point prompts me to imagine a super-scientific mugger who accosts me as follows: "Hand me your money, or I'll make you much smarter, more athletic, and longer-lived!" It's not much of a threat.

Similarly, the word "deserve" might not seem apt if I ceased to be human by gaining more-than-human abilities. If the word "human" ceased to be descriptive of me because I'd become superhuman or posthuman or divine, this would not obviously represent a lost honour or prize. Perhaps there is something arrogant, hubristic, or otherwise morally questionable about aspiring to become superhuman or posthuman – and even more so about aspiring to divinity. But if we actually *became* any of these things, the fact that we were no longer (merely) human would not obviously be a loss. In the upshot, Agar needs to demonstrate not only that radical enhancement could make us, in some sense, no longer human but also that this is a prospect with which we could meaningfully be *threatened*, and something that would make us deserving of a *less* illustrious description than "human".

To be fair, much of the content of Agar's books purportedly demonstrates his points. Does it withstand scrutiny?

2. Humanity and human nature

Insofar as he worries that we might no longer deserve the adjective "human" if we change ourselves radically, Agar owes us some account of what it is to be human. He provides a straightforward definition, or so it appears: "I define humans as members of the biological species *Homo sapiens*" (Agar 2010, 19). But this leads to a difficult topic in the biological sciences and the philosophy of biology. How should we understand the concept of a species, and how should we draw boundary lines between species?

Fortunately, Agar's approach falls within the mainstream of current academic thinking, in that he defines species in terms of populations (or groups of populations) "comprising individuals capable of breeding with each other while lacking the capacity to breed with members of other biological groups" (Agar 2010, 19). As he acknowledges, there is some scientific controversy about the concept of a species, but he is invoking the most common conception of species among contemporary biologists, what is known as *the biological species concept.* This has its origin in the work of Ernst Mayr, and a similar approach is endorsed by Jerry A. Coyne and H. Allen Orr in their authoritative textbook on speciation. Citing Mayr, they define species as "groups of interbreeding natural populations that are reproductively isolated from other such groups" (Coyne and Orr 2004, 27).

Agar argues that "there's a good chance that radical enhancement will in fact create reproductive barriers, and therefore will result in beings who are not human" (Agar 2010, 21). There are obvious problems here in the sense that many human beings are already reproductively isolated, whether by castration, infertility, the outcome of menopause, social isolation, or choice. However, Agar argues that these problems can be solved when we look at the detail. In such cases, reproductive isolation is not total, since "other human beings are the vehicles of an individual's evolutionary prospects", as with a postmenopausal woman who has already reproduced and can now help her children to have their own children (Agar 2010, 23). I accept that these minor issues should not detain us. These problems are no more serious in principle than those relating to any other biologically or socially infertile members of a sexually reproducing species.

Agar is most interested in the *psychological* barriers to reproduction between ordinary humans and individuals who might possess, for example, immensely enhanced intellects or greatly extended lifespans. But he then has to explain how these barriers are more significant than those preventing, say, a castrated man or a postmenopausal woman from reproducing. He responds by emphasizing the *radical* nature of what is involved in radical forms of enhancement:

> There are certainly no hard and fast lines, but radical enhancement is properly viewed as significantly different from menopause and castration. Radical enhancement is more likely than either of these events to result in beings who represent some kind of new beginning. (Agar 2010, 32)

Some additional, more theoretical, problems lurk here. First, the main emphasis when scientists employ the biological species concept is on barriers created by genetic differences, whereas unenhanced or moderately enhanced humans might be distinguished from radically enhanced humans by non-genetic differences such as the use of pharmaceuticals, implantation of cybernetic devices, or systematic and regular use of anti-ageing procedures. Second, the point of the biological species concept is to explain the clustering of sexually reproducing species, coexisting in the wild, into discontinuous groups. It is stretching the concept – and its original purpose – if we attempt to apply it to sub-populations of *Homo sapiens* coexisting within technological civilizations, and reproductively separated, if at all, largely by cultural differences and personal choices. These problems may not be insurmountable, but they at least create a cloud of doubt.

Earlier in *Humanity's End*, Agar observes that "it does seem at least logically possible for a being to begin its existence as a member of a particular species and to exit that species" (Agar 2010, 23). Again, I am prepared to go along with Agar and accept this as a possibility. Change does not, *ipso facto*, entail destruction, so perhaps I could change in a way that does not end my existence as the same individual but takes me out of the reproducing population of *Homo sapiens*. This might happen by incremental modifications to my body over a long time, or perhaps it could be effected more quickly through the use of a powerful, as-yet-unknown technology.

All that granted, the real issue in Agar's critique does not, in fact, seem to be whether we might cease to be human if we employed the technologies of radical enhancement. On occasion, Agar employs *staying human* as a value or goal in itself, but his more persuasive arguments relate directly to whether or not we'd change for the worse if we employed the technologies of radical enhancement, irrespective of whether this caused a speciation event. For example, his main criticism of radical life extension technology is not, in the end, that it might, or would, create a reproductive barrier between those who have and those who have not made use of it. Rather, as we'll see in Chapters Six and Eight, he is worried about such things as whether radical longevity would make us more fearful of death, undermine our current attachments to loved ones and favoured causes, or threaten our ongoing personal identities. In principle, all of these could be consequences of radical longevity even if radically long-lived people continued to have sex – and babies – with unenhanced human beings. Likewise, possible conflicts at the social level could arise between radically enhanced humans and unenhanced (or moderately enhanced) humans even if substantial interbreeding continued.

Thus, I don't deny that some radical bodily modifications might make a person no longer a member of the species *Homo sapiens*. However, this possibility does little of the heavy lifting within Agar's critique. A more promising approach that Agar might have taken would be to ignore the biological species concept, and perhaps even the whole idea of a biological species, and instead to develop an idea of human nature partly detached from concepts that are useful to biologists.[10]

On this alternative approach, he could focus on a conception of human nature defined by widespread and important characteristics that help explain our behaviour and how we differ from non-human animals. Allen Buchanan has sketched an approach somewhat along these lines, specifying, in addition, that the characteristics concerned belong to "*most beings that are uncontroversially human*", both now and throughout human history, and that they "*are relatively recalcitrant to being expunged or significantly altered by education, training, or indoctrination*" (Buchanan 2009, 353; Buchanan's italics). As Buchanan explains, such an approach does not require that human nature be fixed or that the relevant characteristics be strictly universal.

Using this approach, Buchanan is able to explain a sense in which a series of enhancements could produce beings whom we'd regard as posthumans. They would have a different nature from ours, although they might coexist with us in the same societies unless conflict led to their, or more likely our, extinction (see Buchanan 2009, 353–354). Although Buchanan does not make this point (and for all I know might not agree with it), biologists might still classify at least some of these posthumans as members of the species *Homo sapiens*. Their categorization as posthumans would be largely intuitive, rather than scientific, and might depend on differences in a combination of appearance, behaviour, and abilities.

This is not Agar's approach, but perhaps something like it is implicit in his thinking and supplying a source of his concerns. He does, at least, seem to believe that there is a universal, or near-universal, set of values that is currently shared by human beings, but which might not be shared by people who have been radically enhanced. *Pace* Agar, these people might still be *Homo sapiens*, even if certain important characteristics and values have sufficiently changed that they are also, in a different sense, *posthuman*. When I use the expression "posthuman" and its cognates from this point onwards, I have in mind beings who have deviated in important ways from historical human characteristics as a result of radical enhancement to their capacities,

10 To be clear, Agar is well aware of this approach. He rejects it for being too vague and leading to intractable philosophical disputes (Agar 2010, 19–20).

but who might or might not still be members of the biological species *Homo sapiens*. That, I believe, is consistent with the use of this terminology by contemporary transhumanists, who differ from Agar in that they actively advocate a transition from human (in something like Buchanan's sense) to posthuman.

Even this approach might not be adequate for all of Agar's purposes. It would explain how there could be coexisting groups of humans and posthumans, even if they would not necessarily be classified by biologists as separate species, and why they might come into conflict. They might differ from each other in visible ways, analogous to "racial" differences between human populations, such as skin colour and certain facial features. They might also differ in their values, ideals, and practices, in ways analogous to the differences between human cultures. Furthermore, they might differ in capability and threat advantage, in ways analogous to differences between nations or civilizations employing different technologies. All of this might generate conflict, and I'll return to that worry in later chapters. However, most of Agar's criticisms of radical enhancement do not involve fears of conflict between rival populations of posthumans and unenhanced (or moderately enhanced) humans.

For example, no such conflict is relevant to Agar's case that radical life extension would increase our fear of death. What is relevant, perhaps, is that undergoing radical life extension might change our values in such a way that we no longer view an increased fear of death as all that bad. We might come to see it as a manageable problem. Agar would view this as a change from holding human values to holding posthuman values, but another approach would categorize it merely as a change from our current values to a new set of values following a transformative experience. As will be discussed in Chapter Four, similarly dramatic changes of values are already commonplace as a result of familiar experiences such as education and parenthood.

3. A first look at Agar on radical life extension

In Chapter Six of *Humanity's End*, Agar commences a full-scale attack on the alleged benefits of negligible senescence or radical longevity. He concedes that, other things being equal, it seems better to have a longer than a shorter life, that it is better to die in, say, our eighties than our twenties. However, he doubts that this scales up to very long lives far beyond those that we currently observe. To make his point, he again invokes the questionable concept of a *human* life:

> But this preference for longer rather than shorter may not apply outside of the limits of normal human life spans. It could be rational to desire that the events of birth and death frame an on average shorter but recognizably human life. (Agar 2010, 110)

Agar explains, quite correctly, that an extremely long life, perhaps of the order of hundreds or thousands of years, would be a deeply transformed one. We should not think of it as just like our current lives plus some added time to spend on the activities that we find enjoyable or meaningful. Agar compares people who reject the opportunity for radical life extension to mountaineers and others who enjoy dangerous pastimes, in that they don't want to die but are prepared to take risks in order to live in a certain way. Says Agar, of people who would reject radical life extension, "They want to continue living what they consider to be a recognizably human life" (Agar 2010, 111). On the same page, he observes that we would not choose to live longer by becoming Galapagos tortoises.

It's true that the more dramatically a human lifespan is extended from the current average or the historical maximum, the more it will be transformed. However, no one involved in debates over radical enhancement suggests that we extend our lives by becoming more like Galapagos tortoises. Agar himself acknowledges that "[t]his example is slightly unfair", since "[t]ransformation into a tortoise would involve substantial cognitive diminishment with obvious consequences for our values" (Agar 2010, 112). The contrast with extended *human* life is clear: becoming a long-lived tortoise would entail living for an extremely long time but without human-level rationality, curiosity, creativity, capacity for deep relationships, understanding of the world, and of ourselves as beings existing in time, and so on. While these are some of the things that we value about being *Homo sapiens*, rather than tortoises or any other kind of animal that we know of, there is no good reason to think that we would lose them by becoming much longer lived or even by transforming into a new species.

It's one thing, then, to acknowledge that not all transformations are desirable even if they add to our longevity. It's another to imply that the transformations produced by, or needed for, radical life extension would take away the characteristics that most significantly distinguish typical human lives from those of other animals.

For different reasons, the comparison with mountaineers is also unfair. Mountaineers take precautions, since they don't actually want to die, but they also take risks by participating in an inherently dangerous activity. Presumably the risks are worth it, from their viewpoint, since mountaineering brings them significant pleasures and a sense of meaning. I am not a mountaineer, but I can imagine that it offers – perhaps among much else – exhilaration, challenge, a sense of discipline and achievement, and the sheer enjoyment of nature's grandeur. For someone who treasures its particular combination of pleasures, mountaineering is worth the risk of injury or death that remains after precautions are taken.

But how is a decision to take up, and then stay involved in, mountaineering supposed to be analogous to a decision to reject radical life extension? Agar suggests that people who reject radical life extension are choosing to continue what they see as a recognizably human life, but that should not be the analogy. The question should be, first, whether they are choosing to continue with a life that provides pleasures and a sense of meaning analogous to those enjoyed by dedicated mountaineers. Second, would they lose such pleasures, and this sense of meaning, if they decided the other way?

These questions invite thought. Perhaps a typical human life in an affluent country brings so much pleasure and sense of meaning that we should stick with such lives as they are, despite the allure of living to the fabulously impressive ages mentioned by advocates of life-extension technology. That, however, needs to be argued, and the question of whether one or another kind of life is best described as "human" in some sense is beside the point. Some lives that Agar does not regard as human *might*, indeed, be worse than human ones, when judged in accordance with values that strike me, and most of us, as plausible. But other lives that Agar would not regard as human might, for all we know, be *better* than human lives when judged in the same way. We don't really know, but perhaps, for all that Agar has told us by this point of his argument, we should take the chance that a radically extended life might be an improvement.

Although these reflections are inconclusive, it's worth repeating that Agar's distinction between the *human* nature of our current lives and the supposedly *non-human* nature of radically extended lives does little philosophical work. A radically extended life *might* be an undesirable one for us in the same sense as a Galapagos tortoise's life would be – but probably not. Giving up on growing old *might* be like giving up what is valued in mountaineering – but that has not been shown.

4. Scheffler and Kolodny on immortality and a "human" life

Similar issues arise in a recent exchange between the philosophers Samuel Scheffler and Niko Kolodny. Scheffler's 2011–2012 Tanner Lectures were on topics relating to death, immortality, and the significance of humanity's continuation after our individual deaths. Subsequently, the lectures were published in revised and extended form, edited by Kolodny, and with responses to Scheffler from several eminent philosophers including Kolodny himself (Scheffler et al. 2013). Scheffler's central claim is that what he calls "the afterlife" – by which he means the existence of other human beings after our individual deaths, and then into the indefinite future – is more important to us than we normally realize. Indeed, we'd lose confidence in many of our existing values, or simply abandon them, if we did not assume the afterlife in this sense. To take an obvious example, an individual's participation in multi-generational work on a cure for cancer would seem futile to her if she believed that humanity would not long survive her own death.

This is an original and interesting claim, but it is not directly relevant to arguments about radical enhancement. Insofar as I have a quarrel with Scheffler, it relates to his other key claim, or set of claims, to the general effect that our values also depend on the universal and pervasive understanding that our lives are limited in time: "[T]o suspend that understanding would call into question the conditions under which we value our lives and their extension" (Scheffler et al. 2013, 95–96). He adduces a number of considerations to support this claim, though it is worth noting that he hedges somewhat in a footnote where he explains that he's assuming only two options: that of actually dying at some point and that of never, at any time, being in danger of death. As he acknowledges, those are not the only logical possibilities.

It's worth pausing here to note that many philosophical discussions that are relevant to this book relate to the second of Scheffler's two options, which means that the issues don't translate straightforwardly to discussion of radical life extension. People with radically extended lives would not grow old, but they could still be killed in innumerable ways. If true immortality – the relevant option discussed by Scheffler – would be a curse, the same might not apply to radical longevity.

Scheffler then points to the staged nature of a human life (in the sense that a typical human life can be divided into several distinct stages), the importance to life as we know it of various kinds of vulnerability and loss, the apparent incoherence of a life that is both human and truly immortal, and the significance of temporal scarcity to our lives (that is, we do not have unlimited time at our disposal). I'll comment on these arguments in Chapter Six. For now, it's sufficient to state that Scheffler understands a human life as having stages from birth to death, each with its characteristic challenges, satisfactions, and so on. He views this understanding as a human universal, even though there are variations of it among the world's cultures. Furthermore, he states, this is essential to our very idea of a life: "It is essential to our idea of a life that it is temporally bounded, with a beginning, a middle, and an end, and with stages of development defining its normal trajectory" (Scheffler et al. 2013, 100). Here, however, he appears to sense some weakness in the ground beneath him, since he immediately makes a supplementary claim:

> The statement that death is essential to our concept of life is not merely a trivial truth resting on a stipulative definition of "life". It is a substantive observation which reminds us that the aspects of life that we cherish most dearly – love and labor, intimacy and achievement, creativity and humor and all the rest – all have the status of values for us because of their roles in our finite and bounded lives. (Scheffler et al. 2013, 100)

Kolodny finds much of Scheffler's case against immortality unconvincing, as do I. Again, I'll save the detail for Chapter Six. For now, take note that an issue arises between Kolodny and Scheffler as to what counts as a *human* life or as *human* values. First, Kolodny makes a straightforward – and I think incontrovertible – point about the issue of stages in a human life:

> But an infinite life span could involve a progression through stages, even a progression through (an infinite number of) finite stages. Compare the progress of humanity. It seems intelligible that humanity should go on forever, but that doesn't obviously mean that humanity would not progress, change, develop, and so on. (Scheffler et al. 2013, 166)

Yes! An unending life might have stages, even though we are not in a position to identify them in advance. Scheffler replies that it is not the mere existence of stages that is important in human lives. Rather, we understand a human life as involving a number of stages including death, "and [...] this understanding exerts a pervasive influence on the content of our values" (Scheffler et al. 2013, 202). Here there might be an element of truth. Human lives have, to date, been of finite length, and it might be true that the *content* of many widespread sentiments and institutions has been influenced by the shared understanding that death under historical and current conditions has been – and remains – inevitable. If, for whatever reason, we escaped the shadow of mortality, if death were no longer inevitable, many values, and much else, might change.

Scheffler and Kolodny debate how much of what we value in human lives would be possible in the absence of inevitable death, its ever-present danger, and the limited time that it allows us all. This leads to a revealing discussion of what is really at stake in the argument. If mortality is something we should embrace, as Scheffler appears to believe, is this because it is a good thing as judged by a set of objectively correct values, or as judged by human values, or as judged by *our* values (perhaps those of all or most current human beings, or perhaps just those of most people who are likely to read the Scheffler–Kolodny exchange)?

Kolodny mulls over exactly what Scheffler is, or should be, arguing. He suggests that the following claim, as a formulation of Scheffler's thesis, is implausible: "If we never died, we would not live a life or a value-laden life, *because no life* that went on forever would be a life or a value-laden life." Instead, Kolodny suggests, it would be easier to defend this weaker formulation of the claim: if not for our mortality, "we would never live a *human* life, or a life laden with *human* values, because no life that went on forever would be a *human* life or a *human* value-laden life" (Scheffler et al. 2013, 168; Kolodny's emphasis).

But this also seems unsatisfactory, as Kolodny soon demonstrates. It appears to rely on a particular, if merely implied, stipulative definition of what is *human*, but if so why should we be so concerned about being human in this stipulated sense?[11] Kolodny asks why we shouldn't "chafe at being confined to the merely human" (Scheffler et al. 2013, 169). Even if the reformulated claim is true, we might prefer a life that is eternal, and so is not a human life as understood by Scheffler. We might prefer a *different* kind of life with *different* values. Kolodny then suggests that Scheffler's point might be that the values of immortal beings would not be *our* values, which have been formed under the shadow of mortality. In that case, "Perhaps we could wish eternal life for homo sapiens [sic] who have not yet become attached to values such as ours. But could we reasonably wish it for ourselves?" (Scheffler et al. 2013, 169).

On this approach, we might not want to relinquish the kind of life we've become attached to, in exchange for a life in which we would be alienated from our current values. If becoming alienated from our current values is something to avoid – if it is one of our current *dis*values – it follows that we would not choose such a radically different life *from within the value system that we currently have.* Accordingly, Kolodny concludes that the most plausible formulation of Scheffler's claim, one that still gives Scheffler (most of) what he wants, might be: "The lives that we are now in the midst of leading could not be laden with the *values that they actually are* were we never to die, because *those* values presuppose our death" (Scheffler et al. 2013, 169; Kolodny's emphasis).

This claim does, indeed, seem more defensible. It's implausible that we would develop the same entire value systems that we have now if we had not grown up knowing that we are mortal. Likewise, it's plausible that our values would change in significant ways if mortality no longer applied to us. Such a change would likely have profound effects on what we do and do not care

11 Although Kolodny does not mention this, the sense implied is clearly not the biological species conception of *Homo sapiens* as a reproductively isolated population of interbreeding individuals.

about. It does not follow, however, that *all* of our values would change. The thought experiment does not assume any changes to our physiology, or to evolved *Homo sapiens* psychology, beyond giving us some kind of shield against death. That being so, many of the same things would still be attractive, pleasant, or beneficial to us, and many of the same things would still be unattractive, unpleasant, or harmful. Nor does it follow – as even Kolodny seems to think – that we could not choose an eternal life from within our current value systems. Although our current values have been shaped, in part, by our mortality and our knowledge of it, they might already include a *resentment* of our mortality. Within our current value systems, our mortality might be a *dis*value.[12]

In the light of Kolodny's critique, Scheffler clarifies and reformulates his position:

> If we never died, then we would not live lives structured by the kinds of values that now structure our own lives or by the kinds of values that have structured the lives of other human beings now and in the past. Moreover, it is at best unclear to what extent we would lead value-structured lives at all. What *is* clear, in any case, is that we would not live anything resembling what we now consider to be a "life". So the fantasy that the lives we are now leading might continue forever is inherently confused and in principle unsatisfiable. (Scheffler et al. 2013, 207; Scheffler's emphasis)

What should we make of this version? Some of it may be correct, though I am suspicious of strong wording such as "inherently" and "in principle" as used here. Scheffler does not seem to earn the right to this language. That said, the idea that we could *eternally* live lives similar to those we are leading now is, indeed, a ludicrous fantasy. But this is because the world constantly changes. It is inconceivable that the daily experience of an immortal person would be similar in a hundred years time – or a thousand, or a million – to that of any current human being.[13] At any rate, there is no likelihood that an immortal human being – in the sense of an immortal member of the species

12 I'll set aside until next chapter the issue of whether our current values really do include the value of holding on to our current values. This does not appear to be an issue between Scheffler and Kolodny.

13 Note, though, that this might have been far less obvious prior to the Industrial Revolution!

Homo sapiens – could go on living a qualitatively similar life into the vastly different future. He or she could, however, conceivably lead a life that is numerically the same, even as it changes. I'll return to this in Chapter Eight.

Imagine, for comparison, one of our Pleistocene ancestors who, miraculously, was blessed with genes protecting her against senescence, and who somehow survived the numerous dangers of life on earth (volcano eruptions, sabre-toothed cats, plagues, invading armies, and so on) until the present day. To do so, she would have been forced to adapt over time to greatly changing circumstances, and we might expect that many of her values would have changed, but she would still be numerically identical to the person she was thousands of years ago. It is contingently and empirically, rather than conceptually, inconceivable that she could live a qualitatively similar life throughout these events and into her distant future.

Remember, Scheffler is not postulating "immortals" who are merely immune to senescence, and who might, as things contingently turn out, live forever. He is postulating, rather, some kind of protection against any danger of death. It's difficult to imagine how beings with something like our physiology could enjoy any such protection, but that's the assumption in the thought experiment. With that clear, it is trivial wordplay to insist that such a person does not have a life. She is a living thing, separate from other living things. It makes perfectly good sense to speak of her "living her life", having "an immortal life", and so on. It's true that such a person will never have a life that is *completed*, and hence open to an overall retrospective judgement by others after she has gone, but it does not follow that she does not have a life at all. Given the facts stipulated, she does have a life. It is an open-ended, continuing life, but nonetheless a life.

Nor has Scheffler given us any good reason to believe that such a person would lack values. She would find reasons to seek out some pleasant, attractive, enjoyable, and beneficial things, to avoid other things, and to rely on still others that could help her efforts at avoidance. She would need to prioritize and to have a hierarchy of goals and desires. In many ways, her overall value system might resemble those of ordinary humans. If she were merely immune to senescence, not totally immune to death in the way that Scheffler is assuming, she would, in addition, need to avoid things that could kill her. But even with total immunity to death, her values might overlap considerably with ours and be structured in a similar way.

We should be content with the weak, defensible claim that – adapting Scheffler's words in his reformulated thesis – the full value systems of immortal human beings would differ, perhaps significantly, from those of other human beings who exist now or existed in the past. Although Scheffler and Kolodny frame their discussion in terms of true immortality, the same conclusion would apply, to a lesser degree, to human beings who are "merely" immune to senescence, and so live indefinitely. They would still be human beings and they would still live value-structured lives, even though they might develop values rather different from ours.

5. Species relativism

In *Humanity's End* and *Truly Human Enhancement*, Agar introduces the ideas of species relativism and the anthropocentric ideal. As far as I can understand, these are essentially the same concept. Agar does not accept *cultural* relativism, defined as the idea that "the truth or falsehood of moral judgments is relative to a culture" (Agar 2010, 13), because he thinks that an investigation of the different circumstances of known societies undermines claims that there are deep inter-cultural disagreements about values. However, he embraces *species* relativism, according to which "certain experiences and ways of existing properly valued by members of one species may lack value for the members of another species" (Agar 2010, 12). Thus, his view is that we should evaluate radical enhancement, or anything else, by whatever can be identified as the universal values of the species *Homo sapiens* (that is, the values of our particular population of interbreeding, sexually reproducing organisms, reproductively isolated from other such populations).

Similarly, Agar defines the anthropocentric ideal, when employed to make evaluations, as a policy of using human standards and values (Agar 2014, 17). He contrasts this with using an objective ideal, understood as making evaluations of enhancement methods by measuring the extent to which an enhancement actually increases our capacities. That, however, seems like the wrong contrast. The contrast should, rather, be between employing values that we possess universally as human beings and employing values that are objectively binding on us, or objectively authoritative, in the nature of things. That is, the contrast should be between universal human values, if such exist, and objective values – if *they* exist.

As an example of the latter, perhaps we are objectively required to maximize universal utility, even though human beings actually prefer maximizing the utility of themselves, as individuals, and their particular loved ones. In addition to universal utility, other possible candidates for objectively authoritative values include persistence over time, complexity, intelligence, and agency (in the sense of capability or power) – values that might be endorsed by many transhumanists. If we made evaluations in accordance with *those* values, we might strongly endorse radical enhancement. After all, radical enhancement could enable us to persist longer and be more capable of affecting the universe around us. Depending on the nature of the enhancement, we might also become considerably more complex and intelligent. For good measure, our happiness might even increase. Values such as these might lead us to pursue a hyperradical agenda, such as attempting to turn the entire universe into a superlatively powerful and complex superintelligence. Something like this seems to be the goal of certain transhumanists and similar thinkers.[14]

For Agar, human morality is ultimately based on deep, commonly held values that are universally endorsed by human beings, at least after rational reflection. Certain things (ways of life and approaches to living, experiences, social institutions, and so on) might, in this sense, have human-relative value, though they would not be valued by extraterrestrial aliens or posthuman intelligences. Since this is a relativist view, Agar is not arguing that we are *objectively* required to make evaluations in accordance with what he thinks of as human values, and nor does he claim that aliens or posthumans with different values are simply *mistaken.*

With this distinction in place, Agar is able to argue in *Humanity's End* and *Truly Human Enhancement* that some enhancements of our capacities nonetheless leave us with less of what human beings actually value. The im-

14 See, for example, Ray Kurzweil's *The Singularity is Near* (Kurzweil 2005, 29). Kurzweil does not, as far as I can establish, refer to himself as a transhumanist, but such sentiments are common in transhumanist circles.

plication is that rational reflection should lead us, as human beings, to reject those enhancements.

In response, I agree with Agar that there are no objective values, waiting to be discovered, by which we can judge the values actually held by particular individuals, societies, or species. I cannot understand how such objective values could ever be identified, and the idea that certain values are objectively authoritative in this way is mysterious.[15] As we've seen, Agar's alternative is to make evaluations using what he thinks of as universal human values. However, just as it's doubtful that objective values exist, it is also doubtful that these universal human values exist, at least in a sufficiently rich and structured form to enable all of us to reach agreement on topics that lie outside human experience to date.

To be sure, *Homo sapiens* is a species with an evolutionary history. As a result, we have an evolved psychological as well as physical make-up, though it remains an open question – and sometimes a topic of heated debate – just how inflexible or otherwise human psychology might be. In any event, I grant that the values of human societies and individuals who live within them successfully cannot take just *any* form. They are bounded and shaped by typical human capacities, both physical and cognitive, and by what appear to be universal psychological tendencies such as love for our kin – especially our own children – and a certain capacity for wider empathy beyond our kin or our local tribe. Such characteristics both facilitate and constrain the moral systems that human societies invariably develop, and which enable human life to be safer, longer, and more harmonious than it would be in a Hobbesian state of nature with no locally recognized limits on what conduct is permitted.

It follows that we should expect *some* commonality in human moral systems and *some* commonality of human values more generally. It does not, however, follow that we should expect human beings from different societies to find that they share the same full system of values after they engage in rational reflection on the matter, or even that human beings from *the same* society would find that, at bottom, they share all the same values. On the

15 For the full story see my earlier book *The Mystery of Moral Authority* (Blackford 2016, esp. chap. 1; for a follow-up discussion, compare Blackford 2019).

contrary, experience shows a wide of array of differences between societies, and often between various individuals, sub-cultures, and moral tribes (Greene 2013) within the same culture.

Differences in values can be very marked between people who have grown up in Western liberal democracies, which tolerate a wide range of rival worldviews and theories of the good. For example, I am confident that some transhumanists sincerely and deeply value such things as complexity, agency, and intelligence, and that they are unlikely to change their minds about this after rational reflection. At the same time, I'm confident that Agar, along with many others, does not value these things in the same way and that he will never do so. Human societies would not be possible if *all* of our values clashed beyond the possibility of mutual agreement or at least compromise. But everyday experience of modern democratic politics suggests that *some* clashes of values are, indeed, fundamental and permanent.

That being so, Agar oversimplifies in distinguishing between objective values and universal human values. In many cases, we can only act, make evaluations, and so on, on the basis of *our own* values, irrespective of whether these are shared by others. Accordingly, discussion over moral and political issues can often go no further than appealing to values that we hope are held in common between ourselves, as individuals, and our immediate interlocutors. In the remainder of this book, I will often refer to "our values", but I intend no more than the values that I expect might be held by many people who read books like this. There is no assumption that we all agree fully, let alone that all human beings share the same values, though I again grant to Agar that all of our values have been shaped to *some* extent by evolution.

We also need to make another distinction. There is a difference between *human values* and *the retention of our humanity* (or of certain human characteristics) *as one of these values.* For all we know, one universal human value – something that is wanted by human beings across all societies – might be transcendence of our human limitations. From within our human values, or simply "our values", we might want to become more capable, intelligent, long-lived, complex, and so on – the values seemingly endorsed by many transhumanists – and the implication might be that we value losing many of the limitations of physical ability, cognition, and longevity that are characteristic of our species. If, in practice, acting fully on those values leads to our becoming creatures of a different species, so be it.

Agar implicitly acknowledges this distinction when he writes, in *Humanity's End*, that such things as lengthened lives and augmented physical and cognitive capacities may fall within human values. As he states, they "are certainly the objects of human desires" and not mere ideological constructs such as transhumanist theorists might create (Agar 2010, 14). Yet, Agar often writes as if one of our values – or one of the universal human values – is simply remaining human. While this does appear to be part of his own value system, it can't be assumed that he holds it in common with his various opponents and interlocutors.

6. I don't want to be a Cyberman

With that in mind, we can give brief consideration to one of Agar's examples of what could go wrong with radical enhancement. He refers to the Cybermen from the long-running BBC series *Doctor Who* as one cautionary example. The Cybermen are capable of long life and (according to Agar) intellectual feats, but they have inhibited emotions. This is one conceivable transformation that would give us some things that we – or in Agar's terms, human beings – value while taking away others. If we became Cybermen, we would be improved in some ways, whether judged by human values or simply *our* values, but we'd lose important things.

At one point Agar ventures: "I will argue that radical enhancement is indeed likely to take our humanity from us. The question we must then ask is what is lost along with our humanity" (Agar 2010, 15). To his credit, this makes the needed distinction between human values and our humanity *as* one of our values. What would be lost if we became Cybermen, in addition to our membership in the species *Homo sapiens*, is much of what we most value in our lives. We would not want a certain degree, or any degree, of longevity and cognitive enhancement if it came at the expense of losing our capacity for emotion.

Dearest to Agar's heart, then, is the idea that a radically enhanced life would be one of diminished value by universal human standards. I believe that he would be on stronger ground to rest his case *on his own values and standards* and to hope that they are, when elucidated and clarified, widely enough shared by his readers to be "ours". This, however, prompts a further question. Even if no objective values exist, why should we apply human values, or simply our present values, rather than whatever values we would develop *after* being radically enhanced? After all, judged against the values held by Cybermen, it's better to be a Cyberman than a human.

The question is not meant to be rhetorical. I believe we have no choice but to apply our current values, perhaps as clarified by rational reflection. But our own values might be confused, and in any event it might be difficult to apply them in the absence of good information about the consequences of our choices. Such problems bring me, in the next chapter, to the topic of transformative choice.

Chapter Four : Transformative Choices

1. Transformations

L. A. Paul's 2014 book *Transformative Experience* uses the fictional example of choosing whether or not to become a vampire. This might have certain advantages, and you might be glad to have done it once it's happened, but no one could know in advance what it would be like. Even if you're given information by vampires who were once mortal humans, and even if they're trustworthy (which can't be guaranteed), you might not truly grasp what they are trying to tell you. Paul calls learning of this kind, which is possible only by first undergoing a transformation, *epistemic transformation* (Paul 2014, 10).

In addition, some experiences are *personally transformative* (Paul 2014, 16). That is, they change us in ways that go beyond giving us new knowledge, such as what a particular fruit tastes like or even what it feels like to be a vampire. These experiences transform our preferences, values, and experiences of ourselves, and thus change the standards that we use in retrospect to evaluate whatever transformation we've undergone. Setting aside stories about vampires, real-world examples include religious conversions, becoming a parent, training as a doctor, joining the military in time of war, or becoming an international celebrity (perhaps by winning a major sports event).

Paul stipulates that by *transformative experiences* she means experiences that are both epistemically and personally transformative. The practical issue, then, is how to make transformative choices: that is, how to decide whether or not to undergo an experience that you expect to be transformative in this sense. Such choices seem to defy rational deliberation, or at least a familiar kind of instrumental rationality, yet transformative choice, or transformative decision-making, is ubiquitous in contemporary social life.

Though Paul is concerned with the experiences of individual people, it is also worth noting that entire societies – or even our species as a whole – can undergo transformative experiences when they are changed in ways that could not be understood in advance, and which alter the personal and social standards by which such changes are retrospectively evaluated. Collective transformations can alter the generality of values existing across an entire society, or even the values that largely prevail among human beings across our planet. Many political upheavals, technological breakthroughs, and other changes have been collectively transformative. Consider changing attitudes to warfare: while wars are still fought for a variety of reasons, entire societies, perhaps amounting to the majority of the world's population, have now renounced glory, military adventure, and conquest as justifications for war. This change in values would have been widely viewed as undesirable in supposedly civilized countries not much more than one hundred years ago, but few people in those countries now regret the change.

Or consider changing attitudes to sexuality. Across a wide range of societies, including all of the European and Anglophone liberal democracies, a vastly wider range of sexual conduct and expression has become socially accepted since the middle decades of the twentieth century. In some circles, especially among highly educated secular liberals, the change was largely complete by the 1970s, following the sexual revolution that began the previous decade. Since then, however, it has extended to a wider range of regions, social classes, churches, and sects. In this case, some people assuredly do regret the change, but this is not the prevailing opinion. And yet, we can be confident that this change would have been viewed very widely as undesirable only recently in historical time.

2. The problem of transformative choice

As Paul acknowledges, epistemic transformation is not always a barrier to rational deliberation. In some cases we can be confident that an experience will be pleasant or beneficial even if we can't accurately imagine what it will be like. In other cases, our current situation might be so bad that almost any attempt to escape it is worth the risk. Conversely, Paul tells us, we can imagine an enormous range of experiences that we know will be horrible even if we cannot imagine how they will feel. For example, we don't know exactly what it feels like to be eaten alive by a shark or to be smashed by a speeding train, but nor do we want to find out! As Paul observes in her discussion of such examples, we can be sure that these are experiences to avoid. More interesting cases are where there seem to be pros and cons to undergoing an experience, but these are understood only vaguely. The *most* interesting cases are where the standards by which we evaluate experiences are likely to be changed by the experience itself. Deliberating about whether or not to have a child is a classic example.

As Paul states, having a child was not very much a matter for deliberation and choice until quite recently:

> Before contraceptive devices were widely available, deliberation didn't play the same role. Often, you just ended up having a child. And to the extent you actively chose to have children, often it was because you needed an heir, or needed more hands to work the farm, or whatever. (Paul 2014, 85)

Under modern conditions, however, much has changed for women with easy access to contraception. In considering the options, women (and their sexual partners) in technologically advanced societies might have access to what seems like relevant scientific research, such as "research that compares life satisfaction and well-being for those who have children to those who do not" (Paul 2014, 85). As it happens, this research lacks a clear scientific consensus, but according to Paul it generally suggests that if you want to maximize your life satisfaction you should probably *not* have children. For parents, the emotional highs may be higher – and the emotional lows lower – than for non-parents, but the general, if inconclusive, tendency from the research is that having children lowers life satisfaction (Paul 2014, 85–87). Still, even if you study this research and talk to friends and loved ones who became parents, you don't really know how *you* would find the experience.

For Paul, it would be disastrous to endorse a standard of rational decision-making that means relying entirely on scientific research of this kind, since it would follow that no one could rationally choose to have children. That is a harsh conclusion to accept, and relying entirely on this body of research, based on studies of the population in general, would mean giving up on considering our own personal beliefs and wishes for our lives, something that Paul regards as giving up our autonomy. She also pours cold water on supplementing scientific reports with our insight into our individual situations and/or advice from people who know us as individuals and also have some insight into us – all of which she sees, plausibly, as being highly unreliable.

I question, however, whether relying on published science involves giving up *autonomy* exactly, since a woman or a couple might adopt this approach voluntarily, without any undue pressure from others. It might be their freely chosen decision procedure, so this is not a threat to autonomy as it is usually understood. Nonetheless, the dilemma remains. Using this research, even if supplemented in various ways, is inadequate as a means of understanding the full ramifications of the choice for you as an individual. Even if most people end up less satisfied with their lives after having children, it doesn't follow that *you* will be. If you decide against having children based on this research, you might be foregoing what would have been a wonderful experience for *you.*

The further twist that makes the example more interesting is that the experience of becoming a parent will almost certainly change your values. Many activities that you now enjoy and would not give up lightly might come to seem too trivial, or too risky, to bother with once you have parenting responsibilities. Other activities that currently sound tedious or repulsive might come to seem surprisingly satisfying and meaningful. Your general attitude to the world might change, in that it might seem like a more dangerous place once you have a young and vulnerable human being in your care. You might become a more anxious, less happy-go-lucky, person, and yet you might not regret the change. Looking back, you might judge your past self as shallow and flippant and your past activities as frivolous. On the other hand, when making the choice before being transformed, you might wonder whether you want to turn into the sort of person who will make *those* sorts of judgements.

In these ways – though I don't suggest in any others! – the everyday decision to become a parent, or not, resembles the fictional decision that we began with about whether or not to become a vampire. It is, by Paul's definition, a transformative choice.

In attempting to solve the problem of transformative choice, Paul introduces variations of the idea that we should ask, in any particular case, "whether we want to discover who we'll become" (Paul 2014, 4), or whether, in the case at hand, "we value the discovery of what it is like to have" the experience

(Paul 2014, 91). However, this seems inadequate. Although it successfully rules out, for example, choosing to be eaten alive by a shark or smashed by a speeding train, it copes poorly with the choice to have, or not have, a child, and with other examples such as enlisting in the armed forces or studying to become a doctor. Paul's idea sounds as if people make, or should make, their decisions to become parents or doctors or soldiers on the basis of curiosity. I'm not sure that this is what she is really trying to convey, but if we take it literally it leads to the dubious claim that the best reason to have a child is to discover what having a child is like. As Rhonda Martens observes in a recent article about transformative choices in the context of enhancement technologies, having a child for the first time for this reason "seems incredibly flippant" (Martens 2019, 9).

Some people might, in fact, be strongly motivated to find out as far as they can what a certain kind of experience is like for those who undergo it. For example, a journalist or a novelist might want to obtain the best possible sense of what it is like to be a busy doctor or a frontline soldier, in order to convey this to her readers. In turn, the readers might want this information for a variety of reasons, including sheer curiosity. It would, however, very rarely be someone's motivation for *becoming* a doctor or frontline soldier – or if it comes to that, a parent. Paul herself discusses difficulties with her approach and how it should be understood, leading her to refer to the character Marlow in Joseph Conrad's classic novel *Heart of Darkness* (1899): "When faced with each of life's transformative choices, you must ask yourself: do I plunge into the unknown jungle of a new self? Or do I stay on the ship?" (Paul 2014, 123).

Paul has not, in my perception, solved the problem of transformative choice. She has, however, provided a vivid account of *why* it's a problem.

3. Transformative choice, radical enhancement, and the fear of death

At least we have some readily available information about what it means to become a doctor, a soldier, or a parent. After all, such people exist. We can observe and talk to them, there are attempts in journalism, autobiography, and psychological fiction to convey their inner experiences, and we have some scientific research – however clumsy in its methodology – on how they experience life. It might be argued that that is enough for rational choice (see, for example, Krishnamurthy 2015 in respect of parenthood). But even if that's so, nothing comparable exists in respect of an individual's choice to be radically enhanced, or the experience of people in a society where radical enhancement has become universal or commonplace. Radical enhancement is probably not like being eaten alive by a shark, but exactly what it *is* like remains unknown.

If ordinary humans were radically enhanced, we can be sure that the experience would be, in Paul's terminology, both epistemically and personally transformative, although this might apply more obviously to some radical enhancements than others.

For example, being free of age-related decline would likely bring a profound change of perspective. Radical cognitive enhancement might quickly bring knowledge and insight into reality that would lead to a dramatic shift of values. Radical augmentation of physical strength would surely be *epistemically* transformative – no one really knows what it would be like to live as a person who is dramatically stronger than all past and current athletes. But would it be *personally* transformative? Would it change our values? That is less clear, but walking the world with greatly augmented capability of any kind might bring a different perspective, and hence changes in values, even before we contemplate what changes extraordinary physical abilities might make to an individual's career and lifestyle. I conclude that this, too, would be a transformative experience by Paul's definition.

In his critique of radical enhancement, Nicholas Agar righly emphasizes its transformative nature. He marshals examples from popular culture, such as the Cybermen from *Doctor Who* that I discussed briefly in the previous chapter, to illustrate the possibility of an undesirable personal transformation. In such cases, the victims of the transformation claim to be glad that it has taken place; for example, the Cybermen, to the extent that they retain emotions at all, are apparently satisfied with their transformations and would not want to revert to being human. And yet, none of the examples on offer –

becoming Cybermen, pod people,[16] or components of the Borg hive mind – seems attractive to us as we are now. These transformations do not strike us as improvements, even if we stipulate that we'd be happier afterwards and would not wish to return to our previous lives. Thus, says Agar, we would be rational to resist transformation into Cybermen or assimilation into the Borg hive mind: "The transformed state is incompatible with the values of the candidate for transformation" (Agar 2014, 14). By contrast, children become adults with different values from those they had as children, but this is a transformation that children look forward to: "Children actively desire to become the kinds of beings that their parents are" (Agar 2014, 15).

It's not clear just what to make of these examples. At a minimum, Agar is correct that we recoil – perhaps rationally – from some kinds of transformation even if we have reason to believe that they would not make us feel subjectively unhappy, and even if we have reason to think that we might be happier after the transformation than we are now or would be in future in our non-transformed state. This suggests that it would be rational for us to reject *some* transformative experiences even if we expected to be happier afterwards – provided only that we would lose something that we currently value more than subjective happiness or contentment. It might, likewise, be rational to resist *some* transformations of an entire society, or of humanity as a whole, even if we believed that the people of the future, after the transformation, would be happy.

But we need to step carefully here, lest we reject valuable transformations out of hand. Not *all* transformations should be resisted, as shown by the example of leaving childhood behind and becoming an adult. Agar claims that radical enhancement would be a negative transformative change, which he defines as one that "changes for the worse the way we evaluate the experiences, beliefs, and achievements that constitute our lives" (Agar 2014, 15). But would it be?

16 From Jack Finney's novel *The Body Snatchers* (1955) and its cinematic adaptations as *Invasion of the Body Snatchers.*

Let's grant that *some* fictional transformations that would alter our values seem unequivocally negative.[17] But that is because we know in advance – or because it is stipulated – that we would change in certain specific ways that would remove things we are deeply opposed to losing, such as our emotions (as Cybermen) or our separateness from other people (as components of the Borg). Given what we know, or what is stipulated, we can be sure in advance that our transformed lives would be seriously worse, judged by the standards that we currently employ given the values that we have. We can liken this to being eaten alive by a shark, not because it will be agonizing but at least in the sense that it is something that we already know we don't want. We do not, however, know this about being radically enhanced. For all we can be sure at this point of the argument, undergoing, say, radical enhancement to our cognition or longevity might be an amazingly positive experience.

One argument to the contrary is Agar's claim, mentioned in the previous chapter, that people with radically extended lives would experience an increased fear of death. In Chapter Six, I will dispute this, but it does seem to be the type of claim that, if correct, might support an argument that radical longevity is something to avoid – analogous to cyber-conversion or Borg assimilation. It involves the potential for a longer life, but in other ways a seriously diminished one. For now, I'll assume Agar's scenario for the sake of argument.

A heightened fear of death would seemingly apply at the individual level and the social level. At the individual level, I would fear death much more than previously, once I'd undergone a process to reverse, halt, or significantly retard the process of age-related decline. I would become more anxious about risky activities, including many that I currently value and enjoy. At the social level, a society of people with the potential to live indefinitely would be far more risk averse than historical and current human societies. Either way, living with a heightened fear of death sounds unattractive, and if we were saddled with such a fear we might find ourselves (and perhaps our entire

17 In addition to Agar's examples, we should include vampires, at least as depicted in Bram Stoker's classic novel *Dracula* (1897).

society) shunning numerous valued activities, possibly even some that currently make our lives seem worth living.

Let's assume, however, that the experience would be personally transformative. If I underwent a process for radical life extension, my values would change. I might be glad that I underwent the process, I might find new, safer activities to enjoy, and I might dismiss the activities that I previously valued as worthless or unimportant. I might be able to manage my fear of death, given that new activities are available to me. Perhaps I'd become happier overall. But this suggests that my post-transformation self might end up living, and embracing, a life that I'd now despise. He might, in turn, despise the way I live now and my current set of guiding values. This seems worrying!

An entire society of life-extended people would also change its values, and it might develop new institutions and popular activities. Its citizens would not want their society to return to what it was, and they would disparage its previous customs and practices. They might congratulate themselves for the high value their society places on their safety, while viewing earlier human societies as aggressive and immature. But what if the people from an earlier stage of this society knew about this in advance? How would they react? Suppose they are considering policy options relating to radical life extension: whether, for example, research into it should be given urgent, Manhattan-Project-style public funding, given some smaller boost in public funding, left mainly in the hands of the private sector, hindered in some way, or even banned. If they know how their society would change from radical life extension, they might see it as a change toward decadence and cowardice, and they might be appalled.

So, given the stipulated scenario, should we choose radical life extension or not? If we accept the thought experiment at face value, with its various stipulations, should we apply our current values or those we would have *after* the life-extension procedure? Can we get closer to solving this kind of problem? This question does, I think, have an answer – but it isn't straightforward. To make progress, first we'll need a detour.

4. The jungle or the ship?

If we follow standard decision theory, we should make decisions by comparing the expected values of each available line of action. Each of these expected values is quantified in accordance with the probabilities and levels of desirability of the possible outcomes of a particular line of action. In turn, evaluations of desirability are made using our current value systems (interpreted broadly to include our full range of desires, preferences, and so on, whether or not they count as values in any narrower sense). On this approach, whenever there's a clash, we should apply our current values, rather than those we'll develop *after* we make and implement our decisions. On my reading of Agar's critique of radical enhancement, this is also his approach to the problem. Indeed, I see no way of stepping out of the totality of our existing value systems when we make choices. When considering whether to undergo radical life extension, I cannot apply any values other than those that I have now, although those values already include the value of applying my *rationally clarified* values rather than (say) my momentary impulses. As a further twist, however, there are psychological limits to how far anyone can do this.

As a general proposition, we can try to improve our deliberations by clarifying our values and by taking care in assessing probabilities. We can, for example, slow down and deliberate about what we really want. We can try to ensure that we're as well-informed as can reasonably be achieved in the circumstances, and that we're not being inconsistent, or illogical, or giving too much weight to impulses of the moment rather than our longer-term goals and desires (see Blackford 2016, 28–30 and the texts referenced there). We might also take steps to be aware of known cognitive biases that could distort our decisions.

Unfortunately, transformative choices are about experiences that would be epistemically transformative, which entails that we know little about the probabilities of various outcomes. Perhaps it would help if we at least knew the worst thing that could happen, and we might decide to avoid it at all costs (the classic maximin solution to choice under uncertainty), but we might not know that. In addition, there are two more difficulties even if it's agreed on all sides that we should apply our present values when we deliberate. First, our values are not entirely transparent to us. Sometimes we have a sense of what we really most wanted only *after* making an intuitive decision that we find we can live with and "own". For example, a decision to reject cyber-conversion might prompt reflection on what deep values prompted it, rather than flowing from conscious application of values that we have already identified and assigned weights.

Second, as is well known, human preferences are complex and multi-leveled (here, the *locus classicus* is Frankfurt 1988). They include specific and relatively short-term preferences (I might have a preference for going to raucous late-night parties while accepting that this will not always be the case) alongside more general, abstract, and long-term preferences that might be more deserving of classification as my *values* in a narrow sense (I might have a long-term preference for a life that I'll experience as satisfying and meaningful). I might have second-order preferences about which preferences I wish to retain, and perhaps about which new preferences I'd like to have if I could (reflection on this might suggest that, sooner or later, I really ought to ditch my love of raucous late-night parties). I might even have a general preference that I be open to changing my preferences – even long-term ones to which I am deeply attached – as I experience life and grow personally.

Thus, making a decision whether or not to have children might compel a potential parent to reflect on her overall value system, and especially about her priorities to the extent that she understands them, and she might be prepared to sacrifice some preferences for the sake of others. After reflection of this kind, she might see the likely, or possible, effect of parenthood on her personality as equivalent to cyber-conversion – something to be absolutely avoided. Or she might see it as a door to adventure and personal growth, more equivalent, at least in her case, to growing up. But – here's the first problem again – the structure of her preferences might not be entirely transparent to her.

In all, it's not surprising that momentous jungle-or-ship choices are made largely by intuition, then given retrospective rationalizations. Standard decision theory is elegant and, as far as I can see, correct. And yet, it does not operate as a conscious decision procedure when we make these choices. Often, we must simply choose the jungle of a new self, or else stay on the ship. We draw on levels of our values that we could not articulate. Insofar as this happens, we engage in self-deception if we believe, after the fact, that we consciously applied standard decision theory or that we're well placed to use it to justify what we did.

In a sense, this is pessimistic. I'm not, however, claiming that all efforts to be aware of biases, poor reasoning and the like are useless. The deeply intuitive nature of many decisions may not be good news, but it's best to be aware that our values are multi-leveled and not entirely transparent to us. This can, at least, give us more insight into our decision-making processes.

5. Human enhancement and status quo bias

When it comes to (moderate or radical) enhancement of human capacities, some biases are likely to be especially relevant and worth identifying. An example that has received much discussion is status quo bias, the subject of a 2006 article by Nick Bostrom and Toby Ord, published in the academic journal *Ethics*. These authors categorize status quo bias as a bias in ethical judgements and political choices. They define it as "an inappropriate (irrational) preference for an option because it preserves the status quo" (Bostrom and Ord 2006, 658).[18] Their main example of the bias's operation is cognitive enhancement, whether viewed as the enhancement of a unitary capacity – "intelligence" – or of "specific cognitive abilities such as verbal fluency, memory, abstract reasoning, social intelligence, spatial cognition, numerical ability, or musical talent" (2006, 656). According to Bostrom and Ord, cognitive enhancement would have social as well as personal benefits:

> Diseases need cures, scientific questions need answers, poverty needs alleviation, and environmental problems need solutions. While a widespread increase in intelligence may not be sufficient to achieve all these goals, it could clearly help. (Bostrom and Ord 2006, 660)

---

18 For a thoroughgoing critique of the idea that status quo bias is always irrational, see Nebel 2015. Nebel raises deep issues, but as far I can determine, his analysis does not affect the examples that I discuss in the text. He does not claim that the considerations he adduces entirely save the rationality of resisting moderate enhancement of cognition or that they save the rationality of giving false rationalizations of such resistance.

Yet bioconservative critics of enhancement technologies put forward a wide range of reasons, at least some of them apparently spurious, to deny that cognitive enhancement would be beneficial in aggregate. For example, they claim that the only benefits would be those accruing to individuals in social competition with others. In response, Bostrom and Ord speculate – I think plausibly – that most of the critics who suggest that widespread cognitive enhancement would be worse than the status quo also think, or would be likely to think, that widespread *reductions* to human intelligence would be worse than the status quo. But if they think both of these things, they seem committed to the idea that human intelligence is currently at a local optimum, "with small changes in either direction producing something worse" (Bostrom and Ord 2006, 664). This seems unlikely – or at least, anyone who seriously makes such a claim should provide some justification for it.

Bostrom and Ord grant that such requests are not always impossible to meet, but they examine and criticize some general ways to reply "in the case of medically safe, financially affordable cognitive enhancement" (Bostrom and Ord 2006, 665). In brief, they conclude that some human parameters, such as heart-to-body-size ratio, may, indeed, be at an optimal point for human well-being when everything else is held constant. But they see intelligence as a poor candidate for optimality in this sense. This seems intuitively reasonable, at least when considering *moderate* levels of cognitive enhancement.

Next, Bostrom and Ord discuss possible transition costs from cognitive enhancement. They claim that these are easily exaggerated, and they note that the benefits would be permanent while transition costs would be temporary. Among the costs they identify are the need to modify school curricula, as well as some games and recreational activities, to make them more challenging. Again, society might have to cope with friction between parents and their cognitively enhanced children, assuming that the enhancement is done by a method such as genetic engineering of the upcoming generation. The authors suggest that costs such as these could be dealt with relatively easily, and that might be true in some scenarios. At this stage, however, I feel the beginnings of doubt.

The problem is that *radical* cognitive enhancement might produce more significant costs, including costs that we cannot readily imagine. Sufficiently extreme levels of cognition might, for all we know, produce troubled and alienated minds. Setting that aside, the upcoming generation might have minds so deeply different from those of existing people as to cause profound difficulties in communication and mutual understanding. Indeed, most parents would surely reject the abrupt introduction of cognitive enhancement to create children *radically* smarter than them. The children would soon come to regard the parents as no more cognitively competent than we view infants or nonhuman primates. As Agar observes, this "would be a pretty terrifying prospect to many parents" (Agar 2010, 191). Opening such a cognitive gulf between people of successive generations would ruin many valuable interactions between parents and their children.[19]

Bostrom and Ord provide a footnote in which they observe that what they call their Reversal Test (Bostrom and Ord 2006, 664–665) will not give definite results if a proposed change in a parameter such as cognitive capacity is very large. Their explanation is that "there is less prima facie implausibility in supposing that a larger interval of parameter values contains a local optimum than a smaller interval does" (Bostrom and Ord 2006, 671).

In the case of cognitive enhancement, we might reasonably think, without being affected by any cognitive bias, that a small negative movement in human intelligence (we could think of this as pushing the overall IQ bell curve to the left by a few points) would have negative social outcomes. If that claim can be justified, it seems plausible that we should look favourably upon a small *positive* movement in human intelligence. That's fine so far. However, the situation appears different for *massive* shifts of the IQ curve. A massive shift to the left, so that an IQ of 50 became the new mean, would appear to be disastrous for any modern society (or perhaps any human society at all). It does not follow, however, that a massive shift to the right caused by an extreme kind of cognitive enhancement, would be enormously beneficial. On the contrary, the psychological and social impacts are uncertain.

19 For scientific reasons that are well understood, there is no serious prospect of creating this kind of intergenerational cognitive gulf using genetic engineering. Even if it were achievable using cybernetic implants or some other technology, it is something that parents would resist strongly. Agar's worry that such a scenario could become actual, and thus alienate parents from their children (Agar 2010, 180, 190–191), appears to be based on a fear of an authoritarian imposition of radical transhumanist values on general populations. At least for now, that is not a realistic fear.

Imagine, if you can, a situation where an IQ of 150 abruptly becomes the new mean in an existing society. Such a huge change would surely lead to unpredictable, perhaps momentous, consequences. What if the shift were even larger, so that *all* of a society's members suddenly had IQs over 150, or levels of cognition even higher? It is not implausible that this society would enjoy many advantages that appear enviable, but it might, for all we know, develop values and ways of life we'd consider deeply unattractive. The bottom line is that Bostrom and Ord have done us a service, but one with limited application to the question of *radical* enhancement. It is useful to remind ourselves of status quo bias when we deliberate over moderate proposals for change. This can act as a corrective to overly cautious decision-making, especially in an area that seems baffling and scary, as with human enhancement. When it comes to more radical proposals, however, it becomes more plausible that the status quo is somewhere in the vicinity of an optimum under local circumstances, and that a large and abrupt change might have a significant downside.

Though Agar sees some value in the Reversal Test, he also suggests that the endowment effect – another psychological phenomenon mentioned by Bostrom and Ord – might be important as an explanation of efforts to preserve the status quo. That is, we resist some changes because they might require us to give up something that we currently have and to which we feel emotionally attached. In the context of enhancement technology, however, Agar confusingly interprets this idea not just as an emotional attachment to familiar possessions but as an attachment to our humanity: "The strength of our emotional bond with our humanity may make it rational to prefer this state of being to some objectively superior posthuman state" (Agar 2010, 139).

It would be more straightforward if Agar simply conjectured that some opponents of enhancement place a sentimental value on the broad level of capability that they currently possess (or see around them). That might be true, but if so how much weight does it deserve? For a start, it won't persuade people who lack the same emotional attachment, and for what it's worth the existence of such people suggests that it is not a universal human value that we must honour. Nor, for other reasons, does it seem like a weighty factor for social deliberations. If little were at stake and all other things were equal, the mere emotional attachment to approximately our current capabilities might be a legitimate, or even prevailing, value. However, we'd expect this sort of attachment to carry little weight in deliberations where there is much else at stake. If we suspected that highly consequential decisions were being strongly influenced by the endowment effect, we'd likely regard this as a factor to try to correct for, if possible, rather than one to be swayed by.

It seems, in all, that personal or social choices to oppose *moderate* forms of enhancement might be based, in part, on biases (such as status quo bias) and values (such as emotional attachment to what we currently have) that it would be rational to identify and resist in order to attain probable and significant benefits. We can keep these points in mind as part of our more general efforts to improve our deliberative processes.

So ends my promised detour.

6. Death, life extension, and the fear of death

As individuals or as participants in our societies and their decision-making processes, we should not take radical enhancement lightly. Personal or social choices to oppose *moderate* forms of enhancement might be based, in part, on biases and values that it would be rational to resist. Where, however, we opt for more extreme levels of enhancement, it is more likely that there's a genuine downside to watch out for. Radical enhancement is *radical*, so its long-term consequences are speculative. Even the probabilities of various shorter-term impacts are not known. That alone might be a reason to hesitate. I don't propose that we simply give up on assessing the consequences, but contemplating radical enhancement produces a kind of vertigo. At the same time, some dire predictions may not come true and some pessimistic arguments may be flawed. In Chapter Five, I'll examine a pair of such arguments.

Meanwhile, after our detour and these reflections, let's return to radical life extension and the fear of death. It might appear from the last past few pages that Agar is straightforwardly correct in the following sense. *If* radical life extension will most likely produce the effects that he predicts (heightened fear of death and all that it would involve), *then* we should judge those effects using our current values, not values that we'd develop after undergoing radically life-extending therapies. That is a tempting, and correct, conclusion as far as it goes, but it fails to reckon with the crucial insight that our current values are complex, multi-leveled, and not entirely transparent to us.

I might value certain specific activities that involve risk: perhaps I love racing motor cars competitively, or perhaps I'm a keen amateur boxer, and in either case I might be horrified at the idea of a life without my favourite pastime. Or I might value simply being able to drive a car on ordinary roads. On Agar's account, radical life extension would require that I give up all of this. At the same time, I might care even more about my future on-balance happiness, irrespective of which particular activities contribute to it, and about the number of years that I might enjoy a life that, again on balance, seems to contain more good than bad. I might even value developing, in the future, whatever new values would conduce to making this possible. Thus, calling on my current values, in all their complex entirety, I might choose radical life extension, even accepting that it will have a downside and that its consequences will include giving up some of my current values and embracing some new ones. It is possible that my assessment of all this, when I'm placed in a real-life situation, would be largely intuitive and unconscious – yet I might come to feel comfortable with it even before acting on my decision.

It appears, therefore, that Agar can defend only a relatively weak claim. For *some* people, radical life extension – assuming it brings a heightened fear of death and all that goes with it – will be equivalent to cyber-conversion or Borg assimilation. It will require that they give up something whose loss they consider fundamental and cannot tolerate. Agar seems to be such a person, and I might be as well (*this*, in fact, is not transparent to me but seems possible). Other people, such as Aubrey de Grey whom we'll meet in Chapter Six, could make a different assessment without thereby being irrational.

Once again, in this particular (hypothetical) situation I *might* make the same decision as Agar. He, and I, and all of us might be unwilling to change our values in certain *specific* and *identifiable* ways. I would recoil, for example, at the idea of changing into a Cyberman or simply into a version of myself that seems, to my imagination, malevolent or corrupt: for example, into a psychopath, a paedophile, or a fascist, or simply into a cynical, venal criminal. But in many situations we each, paradoxically, place a value on our values, as an overall system, being open to change.

7. A coda: On values and transformation

One of the reasons that we seek such things as education and life experience is that they are likely to give us a new, and perhaps wiser, perspective and the revised values that go with it. In these cases, we hope that changes in our values will be more like enlightenment, growth, or liberation than like being corrupted, harmed, or diminished.

Something similar is true at the social level: in liberal democracies, at least, we tend to value a certain openness to social values changing. Change can come from historical experience, inquiry and discussion, and the struggles of rival ideals in numerous ways and settings. Again, we might recoil at some specific, identifiable changes. We would not, I assume, want our society to adopt the values of a fascist state. But it might be just as well if the political classes and the general population do not know in advance what new values will come to prevail in the future as a result of advocacy and events.

Open-minded people and open societies are vulnerable, in that they actively place their value systems at risk. At the social level, some elements of constitutional design are intended as precautions against the worst, but there is no obvious precaution at the individual level. If, however, we dwell on this concern, it might lead us to be defensive in how we lead our lives, and less open to a range of experiences after all. That would be yielding too far in the direction of caution. It could mean missing out on many wonderful and greatly beneficial experiences. The wiser approach is to be open to a broad range of otherwise-attractive experiences. I'm not suggesting that you undergo Borg assimilation, if that's on offer, or that you sign up to be kidnapped and brainwashed by a gang of Nazis. But by and large, our most precious values can take care of themselves.

Chapter Five : Super-Athletes and Super-Scientists

1. Two arguments

One of Nicholas Agar's central claims in *Humanity's End* is that performance-enhancing drugs "alienate us from our sporting heroes" (Agar 2010, 192). Agar extends this to other modifications to enhance athletic performance, and it is a key argument for him that radical enhancement would produce a world that we'd recoil from, given what he sees as our universal human values, partly because of what would happen to sport. In what follows, I propose to concentrate on physical abilities used in sport such as strength, speed, and coordination. However, it's worth taking note that mental abilities, such as the ability to concentrate and the ability to overcome nervousness, are also crucially relevant to sporting performance, and these could be targets for *cognitive* enhancement of athletes.

While Agar sees radical, or even moderate, physical enhancement as ruining sport, he imagines radical *cognitive* enhancement ruining science. In one way, this is a more interesting argument, since there may be little social interest in encouraging radical physical enhancement, but there seems, at first blush, to be much in encouraging radical cognitive enhancement.

In previous eras of our species, improved physical abilities would have been useful for hunting and fighting, but that is now almost irrelevant. Perhaps there could be military applications for radical physical enhancement, but even here the direction of modern warfare is toward sophisticated cybernetic weapons, with little call for enormously stronger, faster, better coordinated soldiers, sailors, and airmen. Likewise, firefighters do not need to be extraordinary athletes, though high levels of strength, endurance, physical coordination, and so on are doubtless helpful for carrying out their important jobs. It's conceivable that radical physical enhancement would be useful for space exploration,[20] but under current and foreseeable conditions the most likely use would be to produce more impressive and successful athletes. The benefits would primarily accrue to specific participants in sporting competitions, and if they were widely available they would tend to cancel each other out.

The individual benefits of physical enhancement might not be *entirely* positional, and there are probably still many occupations and recreations where above-average physical strength is useful. I doubt, however, that there is much need for either individuals or their societies to encourage *radical* enhancement of such characteristics. By contrast, radical cognitive enhancement is a more immediately attractive prospect when viewed at the level of the individual, a society, or humanity as a whole. It would seem to provide benefits to individuals that go far beyond positional advantage in social competition with others. It could bring a more complex and accurate understanding of the world for the purposes of our daily lives, and it could contribute to the progress of science.

Given these considerations, radical physical enhancement might not ultimately be an attractive option even to transhumanists and other pro-enhancement theorists. Nonetheless, Agar places considerable emphasis on the harm that is caused by physical enhancement in sport, so I'll turn first to this issue.

2. Agar on enhancement in sport

Agar's main concern when it comes to physical enhancement is its effects, as he sees them, on sports institutions as they currently exist. His argument seems to assume a scenario in the present day or the very near future, with most people not yet employing enhancement technologies to any significant degree. He does not address how sport might work in a more futuristic

20 The possible uses of enhancement technologies to deal with the rigours of space exploration could be a topic for an entire book. For now, I doubt that radical enhancement of gross physical abilities such as sheer strength would be useful, but this is certainly an area where I could be wrong.

world, after a period of transition, where, for whatever reasons (perhaps more use than I expect in military contexts or in space exploration), radical physical enhancement has become commonplace.

In dealing with his chosen scenario, Agar contrasts an interest in watching extremes of performance with watching performances with which we can identify. Of the latter, he states: "They should be sufficiently good that we enjoy watching them. But they should not be so good that they are well beyond what beings like us can do" (Agar 2010, 193).

In support, he develops the view that we are drawn by sport to see competitors pushing up against human limits. He evidently has in mind elite international sport, since the participants in most local sport go nowhere near human limits in their performances. On this account, a different species with different capacities might care little about human performances that we find impressive, whereas we might not care much about what *they* can do. Hence, he observes: "We would rather watch Usain Bolt run 100 metres in 9.58 seconds than see athletically superior posthumans or aliens run that distance much more quickly" (Agar 2010, 194).

Indeed, if radically superior posthumans suddenly appeared and competed against current human athletes, the result might not be all that interesting once the initial novelty wore off. For comparison, we are not usually interested in seeing how well humans run against thoroughbred horses, or how long a martial artist can last in unarmed combat against a Siberian tiger. Humans do race horses against each other, although we give much of the credit to jockeys and trainers, and we also race greyhounds and some other animals. In general, however, sport is a contest of physical skills between human beings.[21]

In *Humanity's End* and *Truly Human Enhancement*, Agar invokes simulation theory to explain this, by which he means that we simulate what we take to be the mental states of the contestants, much as we might simulate the inner experiences of characters in a psychologically realistic novel. We cannot meaningfully do this with beings whose mental processes are significantly different from our own, as we might expect with computers, intelli-

21 Further, it is between *Homo sapiens* whose characteristics have not been altered in some way that makes them seem, intuitively, posthuman.

gent aliens, or people who've undergone sufficiently radical cognitive enhancement. When we simulate the experience of a great football goal, a sensational backhand in tennis, or some other impressive act in sport, we experience a feeling of exhilaration that we take to be something like that experienced by the athlete. As an example, Agar describes a remarkable goal scored by Diego Maradona in a FIFA World Cup match in 1986. Fans watching Maradona's brilliant movements might feel, in a small way, something of the player's own exhilaration as he dodged English defenders and ultimately delivered the ball over the line. Or so Agar suggests.

In *Truly Human Enhancement*, Agar explains this further with the concept of engaging veridically with the performances of athletes. "By veridical engagement," he states, "I mean that we can accurately imagine ourselves doing as the competitors do" (Agar 2014, 36). We engage veridically with the performances of others when we accurately simulate the mental processes behind them, which depends on a high degree of psychological similarity between spectator and performer. Thus, we can engage veridically with the efforts of a human chess grandmaster but not with those of an emotionless chess program.

Agar is correct that we cannot genuinely identify with the machine in a machine-versus-human chess match (or with either of two machines matched against each other). All the same, there's potential for confusion here. At times, the problem posed by Agar seems to be about identifying with creatures that have greatly superior physical prowess, as with the superior fighting capacities of a tiger over a human martial artist. At other times, it seems to be about identifying with beings that simply do not think the way we do – or do not think at all, if *thinking* necessarily includes some kind of consciousness – as with robots and computers.

At still other points in his argument, Agar seems to believe that we are alienated from the performances of steroid abusers in much the same way as from the performances of computers, tigers, and aliens. That, however, is going too far. Illegal performance-enhancing drugs are, by definition, outside the rules of the relevant sport – so those who use them are cheating. But that is a reason to condemn cheats, not to find their mental processes inaccessible. Furthermore, we can easily imagine a scenario in which a drug that improves performance is widely used and condoned, even by everyday people who are not professional athletes, because, unlike anabolic steroids, it has little in the way of nasty side effects. If a particular sport – let's say boxing or tennis, but it doesn't really matter – allowed athletes to take this drug (as everybody else is doing in the community), we'd remain as capable as we are now of identifying with the contestants.

In short, safe and legal *moderate* enhancement does not seem to create a problem for audience identification. For that purpose, we want athletes to be competitive with each other, much like us in overall morphology, and at least *vaguely* like us in the way they think (they should not think like inscrutable aliens from another planet or like computers and robots). This gives us a very broad ability to identify with others' sporting efforts, which is just as well, since it's likely that the best athletes already think differently from most of us, and that the similarities in mentality are indeed vague. They have often mastered techniques that we'd find unintuitive, and the best competitors possess mental skills that I, for one, cannot genuinely imagine.

3. Sport and radical enhancement

Though Agar attempts to make a general case against performance enhancement in sport, his more fundamental goal is to develop a case against *radical* enhancement. He does not focus on this in a concerted way, but it seems true that the introduction of radically enhanced athletes, perhaps with physical characteristics dramatically different from those that are normal in our species (swimmers like dolphins, wrestlers like gigantic bears, and so on), would have dramatic social effects. These are, however, difficult to predict. It would make little sense to pit these clearly posthuman athletes against unenhanced, or even moderately enhanced, humans who'd have no realistic prospect of victory against them. Despite the initial novelty, I doubt that such contests could be sustained on the same commercial basis as current elite sport even with clever innovations in handicapping and the like. Perhaps radically enhanced athletes could be shuffled off to their own separate leagues, but even this might never attract the same interest as the existing leagues for *Homo sapiens* who look more like their audience.

For those reasons, there is little incentive for anyone to make posthuman sporting leagues happen. Still, even if they were introduced, we don't know how much damage it would cause to sport more generally. It might not be all that much. For example, *local* sport would go on for unenhanced humans. As long as the radically enhanced athletes were a small minority, competing against each other in their own leagues, this would allow "ordinary" elite sport to go ahead much as now. For comparison, car racing and horse racing do not prevent running races between humans relying on nothing more than legs, feet, and high-tech sports shoes.

But what about a world more distant in time, one mainly or solely inhabited by people with radical physical enhancements? To get a handle on this, we need to think about the nature of sport and why it is worth preserving. Is the supposed threat only a short-term one, and perhaps only to some aspect of sport as it is currently organized, played, and viewed by spectators, or is it a long-term existential threat to sport itself and whatever is valuable about it? The latter seems unlikely, since radically enhanced people in a posthuman civilization of the future could, and probably would, continue to engage in sporting contests – and we might think that radically enhanced spectators could identify veridically with radically enhanced athletes.

The British philosopher David Papineau defines sport as "any activity whose primary purpose is the exercise of physical skills" (Papineau 2017, 236). He acknowledges that excellence in a sport's skills involves an important mental element: if you do not focus in the correct way, you will not sustain excellent performance. Nor would Papineau deny the importance of tactics and planning, but again these are directed at excellent – and in a competitive situation, superior – physical performance. In all, his definition is broad, especially since it covers many pastimes that are not competitive: it's sufficient if participants want to pursue them for their own sake and develop excellence in them. Accordingly, non-competitive windsurfing, horse riding, rock climbing, and skiing all count as sport for Papineau, provided they are pursued with a certain seriousness of intent.

We might wonder about this analysis, given the immense emphasis in current societies on competitive sport, and on specific team and individual sports where elite competitors can earn huge sums of money for consistently beating the opposition. Furthermore, philosophers of sport do not agree on the definition of whatever it is they are philosophizing about. For example, Stephen Mumford uses a different criterion from Papineau to distinguish between sport and "mere" games: "There need be no difference between a sport and mere games. It is more that the former are endorsed by central institutions" (Mumford 2012, 6). On Mumford's account, it seems that all sport is competitive sport, though he will classify as sport various rather informal recreational activities that, in more formal manifestations, have received official recognition. In a different book, we could consider the most accurate definition of sport, including how broad or narrow the definition should be. But today's elite sporting events qualify as sport under any definition. These events are inherently competitive, involve the development and display of physical skills, and are sanctioned by official bodies of the kind Mumford refers to.

For Papineau, sport is very valuable indeed: "Sport reaches deep into human nature, and can be as important as anything else" (Papineau 2017, 234). This seems difficult to accept if it is intended literally. For example, is a nation's excellence in sport really as important – judged by our current standards of importance, or any other plausible set of standards you like – as its impressive human rights record, its success in handling pandemics and other disease outbreaks, or its ability to avoid famines and respond to natural disasters? Nonetheless, I freely acknowledge the joys and satisfactions of sport – competitive or otherwise – that Papineau vividly evokes. Furthermore, contests of physical skill appear to have taken place in all human societies and have often been accorded great importance. I have no objection, therefore, when Papineau extols the worth of sport as the exercise of physical abilities and the pride of honing and displaying them:

> The real worth of sport, I say, lies in the pure exercise of physical abilities. Pride in physical performance is a deep-seated feature of human nature. Humans hone their physical abilities and take delight in exercising them. Perhaps this originally had its roots in the practical needs of hunting, fishing, and fighting, but we have come to value physical performance as an end in itself. We devote long hours to improving our physical skills, and seek out opportunities to test them. (Papineau 2017, 235–236)

To keep this in proportion, we should bear in mind that not everyone is as enthusiastic as Papineau appears to be about seeking opportunities to test sporting skills. Still, we can grant that what he describes is widespread. Perhaps most of us feel at least *some* interest in honing *some* physical skills, even if they are distant from what is ordinarily thought of as sport.

With that granted, efforts to hone our athleticism and physical skills have now lost much of their significance for human or social survival. A certain level of physical fitness conduces to health and longevity, but no extraordinary athleticism is required for warding off heart disease or cancer. In fact, the truly extraordinary strength, speed, physical coordination, and specialized skills of today's top athletes are honed through equally specialized training and assistance that were never possible prior to the later decades of the twentieth century.

Overall, it looks like a mixed bag when we consider the importance of sport. Still, we have no reason to doubt, and every reason to assume, that radically enhanced people would continue to obtain pleasure from sport in Papineau's broad sense. Even if most of us were physically enhanced to very high levels, I expect we'd continue to enjoy exercising and developing physical skills. The general standard of performance would be much higher than at present, but this need not spell the end of sporting activities. Perhaps those wanting to challenge themselves would need to develop more difficult versions of current sports. At the same time, we could expect that many people would not seek out difficult challenges but merely enjoy the fun of participating in windsurfing, rock climbing, horse riding, or whatever it might be, in the way that most amateurs do now.

But what about *competitive* sport? Again, we have no reason to doubt, and every reason to believe, that radically enhanced people – many of them, at least – would continue to find pleasure in honing their skills at sporting activities and in testing their skills against each other. Once again, the standard in, say, a weekend golf competition at the local club will be much higher if everyday people are physically enhanced to levels dramatically beyond the best current athletes. I have no idea whether golf would remain much the same as now (but with markedly lower scores), remain somewhat the same (albeit with longer fairways, more hazards, and trickier greens), or evolve into something rather different. However this turned out, we could expect that a sport somewhat resembling golf would survive, would continue to be enjoyed by many people, and would provide an outlet for physical competition.

Furthermore, even in a world where everyday people have been physically enhanced beyond the level of the best current athletes, some individuals will be better at sport – and at particular sports – than others. We would not all be enhanced to exactly the same level or be interested in the same things, and some people would probably be more willing than others to dedicate themselves to honing and testing their skills. This would happen in numerous settings, from local sports clubs to international events, and again we have no reason to doubt, and every reason to assume, that many people would continue to feel a thrill from sporting competition, whether as competitors or as spectators of contests between skilled and evenly matched rivals. That being so, organized amateur competitions would continue, as would professional sport in some form.

So, where is the threat to sport from radical physical enhancement? Part of the problem might be that professional sport in its current form – in a hierarchy from local, to national, to international levels – would likely change. It might be organized differently if, for example, different kinds of enhancements were common in different parts of the world. But this does not entail that sport itself would disappear, that *competitive* sport would disappear, or that *professional* sport would disappear. It might mean that new forms of organization would be required, or that some sports would develop differently from place to place, but this might not be such a bad thing if it happened slowly enough to be managed. Even now, the world is not necessarily worse than it might have been because some sports are entrenched in certain countries, or groups of countries, with little interest elsewhere (think of hurling in Ireland, Australian Rules Football, American football, and the relatively small number of countries that play cricket, rugby union, or *jai alai*).

Even if professional sport as we know it faded away, this would not be the end of skilled and competitive amateur sport, perhaps organized at a more local level. In some ways, a world in which sport is organized like that might be an improvement on what we currently have. Here it's worth remembering just how recent an innovation in human history is professional sport as we know it. Nothing like it existed until the second half of the nineteenth century, following the rise of the railways and the urban middle class: "The railways weren't the only stimulus for the sporting boom. As important was the emergence of a mass white-collar middle class with the time and ambition for genteel leisure pursuits" (Papineau 2017, 227). Prior to these developments, sporting activities, organized in any formal way, were restricted to the European aristocracy and to students in private schools and universities.[22] With this in mind, it's not surprising that many sports that remain popular today, including the main football codes, were standardized in the last four decades of the nineteenth century.

Professional sport became a global financial behemoth even more recently. In its current form, therefore, it is the product of a specific stage in the development of human society and technology. While there might be deep human urges to develop physical skills and test them in competition, there have not always been the large-scale corporatized and professionalized sporting spectacles that command attention today. *Homo sapiens* existed for 200,000 to 300,000 years prior to, say, the codification of association football in 1863 or the invention of lawn tennis in the 1870s. Sport might, in a sense, be part of human nature. Professional sport as we currently know it is not.

4. Is radical cognitive enhancement a threat to science?

In *Truly Human Enhancement*, Agar concedes that radical cognitive enhancement might help in obtaining various "external goods", or instrumental goods, produced by science – as opposed to what he sees as the "internal goods" of involvement in scientific practice. That is, radical cognitive enhancement might bring greater knowledge of the natural world and improved control of our environment. These might be very worthwhile if they included, for example, earlier access to a cure for cancer. Agar claims, however, that beyond a certain point there would be a diminishing, or even negative, return on the internal goods of science, such as the inherent goodness of engaging with the world in a spirit of curiosity and wonder:

> Beyond a certain point, greater degrees of objective enhancement tend to bring less valuable internal goods. They satisfy our sense of curiosity less effectively. Another

22 It would detain us too much to compare the various physical contests of classical antiquity such as the original Greek games, Roman gladiatorial contests, and the famous chariot races in Constantinople's Hippodrome. None of these resembled modern elite sport in the way they were organized, though the chariot races of late antiquity were perhaps the nearest equivalent. The original Greek games, including the Olympic Games, were well attended by spectators and highly organized, but participation was basically an aristocratic activity.

> way to put this is to say that our cognitive faculties tend to lose intrinsic value with greater degrees of enhancement. (Agar 2014, 82)

On its face, however, this seems a strange claim. Why should people with greater cognitive abilities than ours find it *less* satisfying than we do to engage in curiosity-driven inquiries into the natural and social worlds? Participation in producing new knowledge doubtless has its own joy, as well as its share of drudgery and frustration, but there is no reason to stipulate that cognitively enhanced scientists would be cut off from this. If anything, everyday human experience suggests the opposite, that we take pleasure in doing things that we are good at. If our cognitive capacities were greater than they currently are, perhaps we would gain (even) greater enjoyment from exercising them. At the least, people with cognitive powers superior to our own could be expected to find the same kind of enjoyment that we do in attempting to understand the universe in which we find ourselves. This could provide them with a sense of meaning or purpose in the same way that it can for us. If some scientists cognitively enhanced themselves, to whatever extent was available to them, they might develop theories that baffle unenhanced human beings, but – at least for all we know – the process would be just as satisfying for *them* as current scientific practice is for unenhanced scientists.

Agar claims that radical enhancement is not even an efficient way to pursue science's instrumental goods, that we'd be better off producing more powerful scientific tools than enhancing our own cognitive capacities. By analogy, stronger workers would not have been much help in building the pyramids more efficiently, whereas better technology for sculpting, hauling, and placing stone blocks undoubtedly would have been.

In fact, we don't know whether, or how far, that is true. Agar's own arguments suggest an assumption that radical cognitive enhancement would enable scientists to produce more sophisticated and powerful theories that more accurately represent reality. Moreover, as with Agar's discussion of radical *physical* enhancement and its effect on sport, it is not clear exactly what scenario he is imagining at any particular point. As we'll see, some of his concerns might have a degree of merit insofar as they apply to a scenario where elite scientists are cognitively enhanced to a radical degree while the remaining *Homo sapiens* in the same societies are not. Although such a scenario cannot be ruled out, it is not, as far as I'm aware, something that advocates of radical enhancement propose. In my experience, they are more likely to advocate the diffusion of increasingly powerful enhancement technologies of all kinds, until radical enhancement is commonplace.

Admittedly, even the most attractive of plausible scenarios might involve a transition period during which elite scientists are among the first to obtain access to enhancement technologies. During that transition, cutting-edge science might be even more baffling to most people than it is now. If, however, the path to radical cognitive enhancement lies through more moderate kinds of cognitive enhancement, each of which undergoes a process of diffusion, the worst-case scenario that Agar appears to have in mind will be avoidable.

Agar argues that the scientific theories developed by cognitively enhanced scientists would involve less simplification of reality than those produced to date by human scientists, but even if this is so it is not clear why that's a problem. One suggestion in *Truly Human Enhancement* is that scientists currently idealize away causal influences that are not important, whereas enhanced scientists would be less likely to do so, and would, in that sense, conduct science in a different manner. But if a causal influence is genuinely unimportant, cognitively enhanced scientists should be able to recognize this: one might think that scientists with radically enhanced cognitive abilities would be *more* intellectually flexible than unenhanced human beings. If so, they should be at least as capable as past and current scientists of, for example, factoring out the effects of air friction on falling bodies – or whatever equivalents might arise in the future.

As a specific example of how science might change for the worse, Agar suggests that radical cognitive enhancement would make it harder for scientists to discover a Theory of Everything, by which he means a unifying theory of physics that integrates our knowledge of all the fundamental forces, including gravity, and makes sense of quantum mechanics and relativity theory in one overarching system of explanation. The suggestion seems to be that it would be more difficult to develop such a theory because standards for accepting it would be raised:

> The more representationally accurate idealizations of radically enhanced science are likely to make something that [the scientists] would accept as an adequate Theory of Everything significantly more complex than a Theory of Everything that uses ide-

> alizations adapted to human cognitive limitations. We may find that our very attempts to dramatically increase the explanatory power of our science increase the difficulty of achieving our explanatory aspirations. Radical cognitive enhancement is likely to be self-defeating if conceived as a means of reducing the overall quantity of scientific mystery. (Agar 2014, 104)

But this makes unwarranted assumptions about the intellectual inflexibility of scientists who would be cognitively superior to ourselves.[23] They could make use of well-known intellectual tools such as idealization to abstract away distracting, confounding details. The presumption should be that scientists who'd undergone a process of radical cognitive enhancement would possess more, rather than less, of the creativity, inspiration, and flexibility of thinking displayed by past geniuses such as Galileo and Einstein. As a reminder, radical cognitive enhancement is, by definition, enhancement to a level of intellectual firepower dramatically exceeding that of such giants.

Perhaps Agar is assuming that the most brilliant of our "ordinary" scientists, aided by powerful computers and advanced observational devices, might succeed in devising a Theory of Everything that meets all their own tests, while not meeting those which enhanced scientists would set. It's likely that scientists with radically enhanced cognition would be more adept at disputing hypotheses and theories, as well as at formulating them, and so they might discard some theories that unenhanced scientists would find satisfactory, or at least would hold onto for a longer time. This, however, sounds like an *improved* rather than defective process of scientific inquiry.

One way to think about it is to carry out the thought experiment in reverse. What if the most intellectually powerful scientists to date had been *less* intelligent than Galileo, Einstein, and their like – though perhaps still of above-average intelligence?[24] Would we expect, in this scenario, that our spe-

23 Though he never says so explicitly, Agar sometimes appears to assume that cognitively enhanced scientists would operate like computers: depending on sheer mathematical processing power to draw their conclusions.

24 This is, of course, a variant of Bostrom and Ord's reasoning discussed in Chapter Four. As indicated in the text, I am not assuming here that radical cognitive enhancement would produce no problems, but it's implausible that a notable problem would be reduced ability to make scientific progress.

cies would have been *more* successful over time in uncovering truths about the natural world and reducing what Agar calls the quantity of scientific mystery? It seems unreasonable to answer in the affirmative. Sufficiently dramatic upgrading of cognition might cause unpredicted problems, and I'll return to this in Chapter Eight and thereafter. But assuming it works as advertised, producing impressively intelligent beings who want to be scientists, we should expect it to augment, not reduce, the capability and authority of science as a source of knowledge.

5. The meaning and continuity of science

It's true that science would change its meaning for human beings if it were handed down to us by godlike aliens. In those circumstances, science would lack continuity with our own past efforts as a species, and it would be baffling and mysterious (to an even greater degree than most of us find it now). Something like this would also apply if sufficiently advanced posthumans developed theories that they provided to our unenhanced selves. But once again, that is only one scenario for the future. The situation is different if most or all human beings, not just elite scientists, have become more cognitively capable. In those circumstances, the practice of science would build on previous human inquiries, with no obvious rupture unless we were all lifted abruptly from ordinary human levels to radically enhanced levels.

Indeed, Agar acknowledges that if *we* were radically enhanced the science produced by radically enhanced scientists would not appear alien and inaccessible to us:

> Our much more intelligent future selves would find its proofs and demonstrations genuinely illuminating. They would genuinely satisfy *their* need to understand the universe and *their* place in it. (Agar 2014, 83; Agar's emphasis)

We could, therefore, continue to obtain science's internal benefits, such as the satisfactions of participating in curiosity-driven inquiry. If our best theories involved less simplification of reality than those produced by science to date, there would be the additional satisfaction of reasonably believing that our theories were getting closer to the actual states of affairs in the universe.

Meanwhile, the value of science lies in its power to produce reliable knowledge. This is valued both for its own sake and "for the guidance it can provide for our understanding and our decisions" (Douglas 2009, 94). And yet, the universe remains baffling. Many scientific problems remain wickedly complex and almost intractable. Our most fundamental physical theories cry out for unification, and it's become a cliché that quantum mechanics, in particular, is open to conflicting interpretations, despite its power to make correct predictions to a very high degree of mathematical precision. Even apart from quantum mechanics, much of what has emerged from twentieth- and twenty-first-century science is already outside the understanding of non-scientists or scientists from other fields. As a result, most of us, most or much of the time, are already in a position where science is, as it were, handed to us from above.

It is unclear whether the widespread availability of radical cognitive enhancement would worsen or ameliorate this situation. It would seem to help if the outcome tended to flatten, rather than to flare out, existing cognitive differences within human populations. In any event, the hyperspecialization, entry barriers, and sheer difficulty of contemporary science are *already* a problem. Widespread cognitive enhancement – initially moderate, but perhaps eventually radical – seems at least as likely to reduce this as to make it worse, although I grant that we don't really know.

Agar also makes much of the satisfaction that scientists take in adding to the scientific narrative from Aristotle to Newton and beyond, including the contributions of Darwin, Curie, and Einstein. The suggestion is that this narrative would be broken if science were carried out in the future by cognitively enhanced researchers. I doubt that any of this is the highest priority for working scientists. Surely what they – and the rest of us – most value about science is its sheer ability to produce knowledge. Nonetheless, allow me to record that I find much of value in studying the history of science, especially in examining which scientific arguments failed, which succeeded, and which contributed to knowledge claims that are now seen as reliable. If the continuity of science were broken and this kind of study became less relevant, that would give me some personal cause for regret.

It is not, however, apparent that radical cognitive enhancement would break the continuity of human science. Even if all scientists of the future, from some point in time onward, were dramatically more intelligent than

Einstein, they would still build on what went before them, including the vast range of observations, evidence, and arguments that have established science in its current form. At any point in time, the body of reliable findings available to cognitively enhanced scientists would be path-dependent, reflecting the history from Aristotle to Einstein and beyond.

Admittedly, the pace of scientific research might accelerate in a world where radical cognitive enhancement was widely available. As a result, the body of findings from the days prior to cognitive enhancement might be dwarfed by new findings. But a similar acceleration in the advance of science has already occurred for a range of reasons that includes the increasing power of scientific instruments and computers, and the professionalization of science as an organized activity. Cognitively enhanced scientists would continue to add to the narrative of scientific progress, even if they caused a further such acceleration or developed more sophisticated theories.

6. Enhancements versus tools

Agar sugests that we should rely on improved tools for the further advancement of science, rather than on enhancing our own cognitive capacities. As we've seen, however, there is little merit in his arguments for this conclusion. First, there is room to pursue both approaches. Second, relying solely on improved tools would do nothing to reduce the existing problem of difficult, hyperspecialized, and seemingly alien science. To the extent that it is possible to address this problem at all, it looks as if widely available cognitive enhancements might have some chance of ameliorating it.

Because Agar prefers improved scientific tools to improved human cognition, he addresses a possible criticism based on the extended mind hypothesis, associated with the philosophers Andy Clark and David J. Chalmers (1998). According to this approach, we already have minds that are extended by various tools and artifacts, ranging from diaries to advanced computers. When we employ these in our deliberations, their contribution is part of our cognitive process, and in that sense we are already cognitively enhanced by the availability of these things. It might, therefore, be claimed that we are already cyborgs. Interestingly, Agar does not reject the thesis that we've become cyborgs through our intimate association with technologies that facilitate human thought. Accepting this thesis for the sake of discussion, he redefines the question as one about what *kinds* of cyborgs we should be:

> We should become the kinds of cyborgs whose new machine parts remain functionally isolated and physically separate from our biological parts. We should become the kinds of cyborgs who access the parts of our minds that contain the Google search engine by way of a keyboard or a touch screen rather than by way of a chip directly attached to our neurons and synapses. (Agar 2014, 53)

Although this move is available as a way of restating the distinction between cognitive enhancement and the use of superior tools, the arguments for being one kind of cyborg rather than another are simply a repackaging of the arguments for preferring tools to cognitive enhancements. Thus, the extended mind hypothesis, if it is true, is not decisive against Agar's position, since he uses its concepts to restate the distinction that he requires. Nor, however, does he claim any assistance from this hypothesis.

7. Looking ahead

Although radically enhanced characteristics such as strength, speed, and physical coordination might have some social benefits and some non-positional benefits for individuals, it appears that they are mainly positional goods and that they would lose much of their point if they became widely available. Accordingly, I do not expect them to receive much social encouragement, and although I have disputed some of Agar's arguments against them I don't ultimately support their use. Accordingly, these sorts of radical enhancements will play no further role in this volume. Current societies have an interest in monitoring and regulating the use of *moderate* kinds of physical enhancement, especially for sporting performance, but that topic has already attracted a huge body of commentary, and again no more need be said about it here.

Agar's claim that radical cognitive enhancement would be harmful to the practice of science, ruining much of what gives science its value for us, strikes me as ingenious but ultimately unconvincing. There are, however, other arguments against radical cognitive enhancement that appear more plausible, including arguments to the effect that it would undermine personal identity. Some of these arguments are worth serious reflection, and they might dampen our enthusiasm for radical cognitive enhancement. I'll return to this in Chapter Eight, where I'll examine such arguments in a general way – not only with respect to radical cognitive enhancement but also as they apply to radical life extension. Meanwhile, Chapter Six covers the issue of radical longevity more generally and introduces some of its friends and critics.

Chapter Six : Radical Life Extension

1. Preliminaries

In *Beyond Therapy*, its thoughtful report on the implications of human enhancement technologies, the President's Council on Bioethics distinguishes what it regards as three broad approaches to extending human life:

> (1) efforts to allow more individuals to live to old age by combating the causes of death among the young and middle-aged; (2) efforts to further extend the lives of those who already live to advanced ages by reducing the incidence and severity of diseases and impairments of the elderly (including muscle and memory loss) or by replacing cells, tissues, and organs damaged over time; and (3) efforts to mitigate or retard the effects of senescence more generally by affecting the general process (or processes) of aging, potentially increasing not only the average but also the maximum human lifespan. (President's Council on Bioethics 2003, 186)

As the authors of *Beyond Therapy* acknowledge, (1) and (2) in this taxonomy are uncontroversial applications of modern biomedical science. The controversy arises with efforts to increase the maximum – and even the typical – human lifespan beyond that enjoyed by anybody to date. In principle, this could lead to dramatic extension of the span of human life, reckoned in hundreds or thousands of years, or even longer, since there is no definable limit to how long we could live if we did not decline with age. These outcomes would fall within the concept of radical life extension.

In recent decades, radical life extension has become more plausible as a scientific goal and social option, which is not to say that it is considered possible by most scientists working in relevant disciplines. What can be said with more confidence is that the idea has gained in credibility, and there is now a live issue as to how much public and private funding should be diverted to life-extension research. For the philosopher John K. Davis, who favours radical life extension, the practical choice is between funding this research as generously as needed "for it to progress as fast as it can" (Davis 2018, 2) and any other, less aggressive, level of funding. According to Davis, prohibition is not a practical option. This appears to be true,[25] though it does seem that we have a range of options ranging from an all-out publicly funded effort to eliminate or "cure"[26] senescence once and for all, through various lower levels of support or encouragement, to a budgetary and regulatory climate of active discouragement.

Writing in 2008, David Boyd Haycock observed a new level of energy and optimism in the field of gerontology:

> Over the last two decades there has been something of a revolution in gerontological research. From once being a bit of a research backwater, ageing has become a hot topic. Change – and, with it, optimism – is definitely in the air. (Haycock 2008, 204)

Haycock elaborates upon this by explaining rapid advances in fields such as cell biology and genetics, and by referencing successful extensions to the longevity and youthfulness of mice, nematode worms, and fruit flies. These extensions of life in a diversity of non-human organisms provoke the question of how far science might be able to extend *human* longevity and youthfulness.[27] More recently, David A. Sinclair, a Harvard-based geneticist and a leading expert on the mechanisms of ageing, has written a book entitled *Lifespan* with the assistance of journalist Matthew D. LaPlante. Sinclair conveys a mood of excitement among researchers on the ageing process, even though they are reticent about expressing it publicly (Sinclair and LaPlante 2019, 214–215). He suspects that many researchers are already taking subs-

25 There are many reasons for this, but one is that research into the mechanisms of ageing cannot be neatly cordoned off from research into age-related diseases. The latter enjoys strong public support. There is scope for considerable expansion of funding for the study of ageing as such, but there is little or no political room to prevent it or to drive it significantly below its current level.

26 There is increasing scientific support for the concept of age-related decline as itself a disease process that it makes sense to try to cure. David A. Sinclair expresses this view straightforwardly: "Aging itself is a disease" (Sinclair and LaPlante 2019, 67).

27 For a more detailed, though still concise, summary of possible methods of life extension, see Davis 2018, 8–15.

tances that are believed to have anti-ageing properties: notably metformin, nicotinamide adenine nucleotide (NAD) boosters, and low doses of rapamycin (Sinclair and LaPlante 2019, 214–215). If so, they are informally experimenting on themselves.

In *Humanity's End*, Nicholas Agar states his confidence in the ability of twenty-first-century biomedical science to retard human ageing to the point where we experience only negligible senescence. For Agar, then, the question is not whether we could produce this result with sufficient effort but "whether we should invest the resources necessary to work out whether and how it can be done" (Agar 2010, 102). Is investment in this form of science a moral and social priority? Might it, on the contrary, be better to discourage such research? On the final page of *Humanity's End*, commenting specifically on the prospect of radically increased lifespans, Agar states, "We may persist, but only with existences that we properly view as impoverished" (Agar 2010, 198). That is the central issue of this chapter.

Radical life extension could bring benefits or costs to entire societies, and to our species as a whole, but my emphasis for now is on the possible impact for individuals. In Chapter Seven, I'll take up the Malthusian issues of population increase, which could produce social conflict and other collective problems, as well as affecting individuals. Likewise, I'll reserve questions of personal identity, individual survival, and so on, until Chapter Eight, since similar questions arise with radical cognitive enhancement – by itself or in tandem with radical life extension. These questions merit a chapter of their own.

2. Senescence, death, and life extension

Though the best definitions of ageing and senescence are debated by scientific experts, as is the evolutionary cause of senescence, most scientifically tenable theories of the ageing process in humans do not view bodily and cognitive deterioration with age as an environmental adaptation (see Davis 2018, 259–263; Sinclair and La Plante 2019, 10–13). Rather, they reflect the reality that life in the wild is ferocious and short for most animal species, including human beings and our evolutionary ancestors. Because the natural environment is so hard, evolution selects for genetic differences that give a creature some kind of reproductive edge while it is still young, and does *not* select for differences that would prolong life in more comfortable, less dangerous environments. As Jonathan Weiner expresses the point, "It is only among the young that variations will be weeded out" (Weiner 2011, 94).

One such scientific approach can be found in Peter Medawar's classic paper "Old Age and Natural Death", first published in 1946 and reprinted in his collection *The Uniqueness of the Individual.* Here, Medawar discusses "natural death" in the sense of death where "a pretty big contribution" comes from senescence, "the age-specific decline of our faculties" (Medawar 1981, 2). In fact, as he explains, natural death in this sense is highly *un*natural. It is an artifact of civilization, which, in turn, can be understood as a kind of self-domestication of *Homo sapiens.* In the evolutionary environment, "under numerically definable conditions" (Medawar 1981, 22), a genetic endowment that favours the survival of younger animals while disfavouring that of their elders will be selected, and will spread through the population. This, according to Medawar, is what has happened to us.

A similar theme can be found in the "disposable soma" concept developed by the British biologist Tom Kirkwood, drawing on ideas from Medawar, J. B. S. Haldane, and George Williams (Haycock 2008, 225–230). Here, the basic idea is that evolution has not favoured the survival of an organism's somatic cells – which make up almost the entirety of our bodies – beyond the time by which the organism would probably be dead by accident or predation. In the wild, evolution will favour mutations that increase growth, robustness, and reproduction, but will not favour mutations that reduce senescence. Wherever there is a trade-off, growth, robustness, and reproduction will win out over resistance to age-related decline.

On these approaches, the senescence of cells and organisms is not nature's wisdom in preventing overcrowding and making room for new organisms. Rather, it is a by-product of evolution acting in dangerous environments. Haycock states: "[O]ur ageing ancestors would not have been worrying about the debilitating diseases facing them in their old age: most of them never got there" (Haycock 2008, 229). In short, senescence did not evolve because it benefits us as individuals or as a species. This undermines the idea that senescence is a valuable feature of human life that we ought to respect and conserve. It does not follow, however, that eliminating or curing it could be done in current circumstances with no undesirable consequences.

Much has been written about those consequences by apologists for mortality, who've had the loudest, most prestigious voices since ancient times, so first it's worth considering another side of the story. The American bioethicist Carl Elliott is not a card carrier for transhumanist thought or prolongevitism, but he has written vividly about the encroachments of old age:

> Often it is not Death that we fear so much as the instruments of torture that he carries with him: the degeneration and disability, the creaking joints and aching bones, the loss of stamina and sexual attractiveness. But what if it were possible to disarm Death – not just to prolong life, but also to extend youth? (Elliott 2003, 276)

Elliott also reports the view of a friend, the age researcher David Gems, who regards the tragic part of growing old not as the approach of death but a form of "ontological diminution" as the conditions of existence flatten or narrow, with our senses dimming, minds slowing, sexual desires waning, and bodies weakening. Then, as the end gets closer, the possibilities for the future seem to close, "not only because we cannot do as much as before, but also because we do not *want* to" (Elliott 2003, 186). It seems perverse to defend this process.

Even if we live to 80 or 90 or 100 years, we function at our peak for only a small proportion of them: roughly our late teens, twenties, and early thirties. Once we live much longer, human life as we currently know it involves constant decline of our capacities. This begins slowly, and we might still be in good shape at 40 – or even 60 with a lot of luck – but eventually the process spirals us down into decrepitude, frailty, and dependence. This impoverishes our lives in basic ways, and it's no wonder that so many people do their best to resist, even when bombarded with slogans about the beauty of ageing, the charm of growing old gracefully, and the like. Against such sickly-sweet rhetoric, it's refreshing to read a strong, concise statement of the expected benefits of extended life:

> Extended life gives you more time to accomplish your life goals and contribute to the world. You would live long enough to see what the future brings and to benefit from the marvels of a more advanced science. You would live alongside the posterity you already care about. A longer life span might give you greater wisdom and maturity, and if enough of us undergo this, the world might be less violent, less prone to war. We might value the future more and take a long-range view when it comes to

> the environment and the welfare of people generations in the future. Finally, and much closer to our present concerns, life extension research may be the most effective way to combat age-related diseases – halt aging and you reduce the incidence of cancer, heart disease, stroke, and diabetes and the suffering they cause. (Davis 2018, 56)

Here, the intrinsic benefits to individuals are obvious, and there would be a straightforward social and economic benefit if the typical pattern of human life in industrially advanced countries were altered, with a higher ratio of robust, healthy years to years of frailty and dependence.[28]

Jonathan Weiner's *Long for this World* is a journalistic investigation of contemporary gerontology and its practitioners. In particular, it focuses on Aubrey de Grey, a tireless advocate for radical life extension whose formal qualifications include a doctorate in biology awarded by the University of Cambridge for a book that he published in 1999. Since then, he has pursued an ambitious project to combat the process of senescence. Though the details need not concern us, de Grey's strategy is basically to fix or "clean up" seven types of things that go wrong – seven types of damage, or "seven deadly things" – as we age. Each of these operates at the level of our cells and their interactions.[29]

Throughout *Long for this World*, Weiner describes his numerous interviews and encounters with de Grey, who comes across as a likeable eccentric, though also as neurotic, otherworldly, single-minded, and rather arrogant. The opening pages of the book recount the detail of a particular conversation between Weiner and de Grey, set in a venerable Cambridge pub. In an expansive mood after a few drinks, and eager to explain himself, de Grey expresses what seems an unbearable sense of responsibility to do all that he can to hasten the abolition of ageing. Overall, the book's pen portrait of de Grey shows touches of affection, but it also paints him as oblivious to most of what Weiner finds valuable in life – and evidently expects his readers to – whether it be history, art, or having children. *Long for this World*, then, is a

28 The latter point is a major theme of Alex Zhavoronkov's *The Ageless Generation* (Zhavoronkov 2013).

29 As summarized by Nicholas Agar, each is "either a problem with cells or with the way cells relate to one another" (Agar 2010, 88).

book (largely) about the world's most militant and high-profile prolongevitist, written by an apologist for mortality who manages to weave in much of the pro-death case.

Spurred by his insight that senescence is the effect of accumulating damage at the cellular level, de Grey has developed strategies to address each form of damage individually: his famous Strategies for the Engineering of Negligible Senescence (or SENS). Unfortunately, while it sounds simple in principle to repair a small number of types of damage accumulating in the body, it is much easier said than done. It means imagining the needed technologies, creating them, testing them, and making them reliable. Furthermore, some of the steps required might be intensely unpleasant. To combat cancer, de Grey has proposed Whole-Body Interdiction of Lengthening Telomeres (WILT), which involves eliminating the enzyme telomerase from our bodies in the hope of eliminating it from cells that might otherwise build tumors. Even if this were effective, it would require other, consequential, steps, since telomerase enables our cells to keep dividing, and is thus vital to us. The raft of therapies associated with WILT might, therefore, be needed repeatedly to mimic the beneficial properties of telomerase, and they might be traumatic to experience even once. WILT would also make us totally dependent on ongoing technological interventions.

De Grey's aggressive advocacy of his program, close associations with the transhumanist movement, and self-promotional skills have given him a high profile, and his work is often a focus for discussion when philosophers and bioethicists consider radical life extension. He is not, however, the only expert who is optimistic about research into age-related decline. On the contrary, optimism, enthusiasm, and excitement about what might be achievable have become more mainstream attitudes in the relevant fields of science. Furthermore, some current lines of research might lead to anti-ageing therapies far less intrusive and traumatic than extreme proposals such as WILT.

3. Mortal and immortal, the shape of lives

Senescence undermines our powers, weakens our desires, narrows our lives, and, if nothing else does first, eventually kills us. Although it would be a difficult task, we might be able to resist senescence through science and technology, and as we've seen some scientists are motivated to try. Should you agitate to give their work a massive injection of public funds? Ending senescence could become another Manhattan Project or Apollo program, an urgent scientific effort enrolling some of the world's finest minds to meet a great challenge.

That, of course, is not the view of apologists for mortality. In the following sections, I'll examine some of the main arguments employed by these apologists, although one complication is that they often present a case against absolute, or true, immortality rather than radical longevity of the kind advocated by de Grey and by many transhumanists. As we saw in Chapter Three, one example is Samuel Scheffler. He assumes that the alternative to mortality is a literally endless life in which there is never a danger of death. The same assumption appears in the relevant work of Martha C. Nussbaum, including a fascinating series of exchanges that she has had with the philosopher John Martin Fischer over the past quarter of a century.

Perhaps this reflects the historical importance of traditional Christian views, which emphasize an afterlife and personal immortality (not necessarily in a good place). Against that background, what I am calling true immortality might seem especially salient to many philosophers, but at this stage of human history it is more helpful to think about such concepts as negligible senescence, retarding or eliminating or curing senescence, or just radical longevity. We can then keep in mind the question of whether any particular argument against true immortality translates well as an argument against radical longevity. With that said, one argument that does appear to translate well relates to the importance of stages in life, including a beginning and an end. We've already seen a version of this from Scheffler, but it's common in the apologist literature.

At one point in *Beyond Therapy*, the President's Council on Bioethics presents a lyrical version of the idea. The President's Council asserts that a human life resembles a theatrical play or a musical symphony, each of which has its conventional and expected stages. The Council denigrates an open-ended life as involving a meaningless homogeneity, or "chronological atomism", of events:

> Some proponents of age-retardation research use language that suggests an image of life as a "time line", uniform and homogeneous, rather than as a forward-moving

> drama, composing different acts or stages – infancy, childhoood, adolescence, coming-of-age, adulthood, parenthood, ripeness, decline. This would imply an understanding of life as composed of interchangeable and essentially identical units of time, rather than composing a whole with a meaningful form of its own, its meaning derived in part from the stages of the life cycle and the fact that we live as links in the chain of generations. (President's Council on Bioethics 2003, 209)

As another example, Weiner refers to the seven stages famously described by Shakespeare's cynical character Jacques in *As You Like It.* Weiner claims that no stage of life would seem meaningful by itself, but each obtains meaning only in relation to the others. As he develops the thought, he concludes: "A huge part of the action and the drama in the seven ages comes from the sense of an ending, the knowledge that all these ages must have an end" (Weiner 2011, 253).

While it is possible to write about the seven, or eight – or whatever the number is – stages of life with a certain beauty achieved through finely honed rhetoric, a human life is *not* a symphony, a stage play, a flower, or a piece of fruit. It is, quite simply, the life of an individual human being, which need not take any conventional form or be aesthetically pleasing to anybody looking on from outside. Under current conditions – those of industrial civilization where people now tend to live for about 80 years – a person's life might go through something a bit like the President's Council's list of eight "acts or stages". In a prehistoric society, however, few babies made it to adulthood, and an experience of the full eight stages was rare. Under other conditions, with more powerful technology, a typical life might have a different shape again, but to assume that it will thereby be shapeless, static, or repetitious shows a lack of imagination. During our youthful days, and even during our robust years of middle age and early old age, we can, if luck and resources permit, enjoy a vast range of different experiences. There is no reason why these should run out as long as we are physically and cognitively capable of living actively. By contrast, senescence diminishes the variety in our lives by gradually removing our capacity for active living. Senescence poisons everything.[30]

---

30 To borrow from the title of a well-known book by Christopher Hitchens (2007).

Toward the end of his article “Epicureanism About Death and Immortality”, John Martin Fischer addresses traditional apologetics for the limited duration of life, dating back to classical antiquity. He questions the ancient claim that human lives have a shape and a limit, much like a banquet, and that we should not try to overstep or overstay the boundaries:

> But the question at issue here is not about our ordinary, normal lives, but about life's possibilities, considered from a distinctively philosophical point of view. Why can a banquet not be a kind of “temporal all-you-can-eat buffet?” Better: why can we not imagine an indefinitely extended banquet, with suitable intervals for recovery (and enjoyment of other activities)? This is, after all, the way “foodies” tend to look at it already! (Fischer 2006, 379–380)

Likewise, Fischer is unimpressed by the argument that our lives require a narrative structure, and hence completion. Why, he asks, can't they have a structure with no end in sight, as with serial narratives such as long-running soap operas? He concludes that we can have narrative elements within our lives – drama, relationships, and so on – without each life forming a completed story. Interestingly, Nussbaum, who was something of an apologist for mortality in the 1990s, now agrees with Fischer that an immortal life could have an interesting and involving narrative shape – something that we could care about and see as valuable. She particularly approves Fischer's idea of an indefinitely extended banquet (Nussbaum 2013, 37).

As Fischer points out, the stage that would be missing for immortals would be one of physical and mental deterioration, with a so-called “second childhood” and a period of preparation for death (Fischer 2020, 114). It is difficult to see why we'd view the lapse into a second childhood and a state of helplessness as something positive, or why we'd miss it. Fischer compares the fairly constant climate of his home in southern California to climates where he's lived in the past, which had more definite seasons. He remarks: “I can't see why I couldn't adjust to a long stage of immortal life in which I am in my 'prime'!” (Fischer 2020, 115). If we want metaphors, life might become a wonderful, unending banquet (as Fischer suggests), an open-ended adventure, or a river flowing into the unknown future.

4. Danger, mortality, and human values

While concerns about the shape of a human life apply equally, or almost so, to true immortals and people whose lives are "only" radically extended, the next argument translates less easily. This is the claim that many experiences, activities, relations, and virtues of character that we currently admire and value would not be relevant to immortal beings. In her classic discussion of this idea in *The Therapy of Desire*, Nussbaum asks us to consider the immortal – but not omnipotent or completely invulnerable – Olympian gods. Since they cannot die, they cannot show courage in the face of death or a willingness to sacrifice their lives. And so, we see in Homer a certain "laxness and lightness" (Nussbaum 1994, 227) in the lives and relationships of the gods, especially when we compare Homer's human characters. Furthermore, many human virtues are not needed by such beings: these are virtues such as moderation, private generosity, and justice in public allocation of resources.

By 2013, Nussbaum had changed her views on this considerably, stating that she had been persuaded by her exchanges with Fischer that "the relevant things are the limit and the struggle, not mortality. One can have plenty of striving and effort in an immortal life, provided that some limits are held in place" (Nussbaum 2013, 37). She now imagines an immortal struggling against many limits, possible setbacks, and sources of frustration and danger – and that these are sufficient to give the virtues their point. Thus, for Nussbaum, a good human life need not involve death, but it must involve elements of struggle:

> I am persuaded too that most of what we value in human love and friendship does not require death, though it does require the possibility of feeling adversity, overcoming difficulties, illness and recovery. (Nussbaum 2013, 37)

Here, I'm inclined to wonder why illness and recovery are so important for what we value in human love and friendship, as long as other dangers and challenges are involved. But in any event, Nussbaum concludes at this point that, however we try to imagine the concrete details of a life for immortals, "we may grant to Fischer that removing mortality still leaves most of what we value in human love and friendship in place" (Nussbaum 2013, 37). Nussbaum still has some caveats, though these are more relevant to true immortality than the prospect of radical longevity. First, the lives of true immortals would be missing a type of intense devotion that involves risking death itself, and, second, they would be missing a type of courage that extends to risking one's life for others. If I were trying to defend true immortality, this would prompt some interesting inquiries, but for present purposes it's sufficient to recall that radical longevity is not a shield against being killed and would not be inconsistent with the forms of devotion and courage that Nussbaum mentions.

As a further caveat, she carefully grants: "[F]or at least some people the prospect of death and awareness of the shortness of life supply a motivation to leave a mark on the world, and this can in at least some cases lead to good actions" (Nussbaum 2013, 38). This might be true, but an indefinitely long life could also lead some people to take on challenges that would never fit into human lives as we know them now.

Although Nussbaum has largely walked away from her analysis in *The Therapy of Desire*, similar ideas remain popular in the community of philosophers. For example, Scheffler claims that our understanding of human life depends on such things as the presence of loss, harm, and danger in our lives, which he sees as largely depending on the certainty of death. He claims, further, that the practice of valuing relies on certain conditions that operate in the background, most prominently "temporal scarcity" (Scheffler et al. 2013, 100). We could not, that is, value some particular things, as opposed to others, if we had unlimited time to enjoy whatever we wanted.

In an exchange with Scheffler that I examined in another context in Chapter Three, Niko Kolodny responds that many of these losses, harms, and dangers would continue to exist for beings otherwise much like ourselves, even if they were guaranteed to live forever:

> There could still be the possibility of loss of honor, reputation, self-respect, good conscience; failing in one's projects; hurting, disappointing, being separated from, rejected by those one cares about; being tyrannized, enslaved, humiliated by others; losing skills, capacities, control over one's body or environment; as well as pain and suffering. (Scheffler et al. 2013, 166–167)

If these sorts of things are important for forming values, even true immortals would have an adequate supply. They would have numerous things to want to avoid, and they could identify numerous things that could help them do so: their own abilities for a start, and the sympathy and assistance of others.[31]

In reply, Scheffler states that we cannot assume that immortal beings would have our concepts of loss, dishonour, and so on, as appearing in Kolodny's list. "More generally," he adds, "we cannot simply assume that the values that figure so centrally in our actual lives would play a comparable role – or any role – in an immortal life" (Scheffler et al. 2013, 204). Notice, however, that this is a weak claim: we cannot *assume* that certain kinds of losses or setbacks to interests would apply in the same way to immortal versions of ourselves as they apply to us. I grant that we should be careful about how much we simply assume, but it seems clear enough that we could, for example, still be humiliated, tyrannized, or enslaved even if we became immortals. These would still be things to avoid. Furthermore, even immortals would be vulnerable to more refined setbacks to their interests such as loss of reputation. As long as they remained social animals, with some reliance on each other to obtain things they wanted and to avoid the things they wanted to avoid, they would need to maintain reputations as willing and competent social cooperators in good standing with their fellows.

This brings me to Scheffler's claim that the practice of valuing relies on temporal scarcity. The thought here seems to be that we cannot value some things over others if there is infinite time available to have all of them at some point. But this is just not true. Even if we had infinite time in this sense, some things would be more pleasant and attractive than others to creatures constituted physically and cognitively like us, some things would fall in the to-be-avoided category, some abilities, resources, and sources of support would help us avoid to-be-avoided things, and so on. There is, moreover, always temporal scarcity even in an unending lifetime. Except for the shadow of mortality, I might have time to spend a year exploring each and every city in the universe, beginning with Rome and Paris, but by the time I'd visited a few thousand others, Rome and Paris would have changed dramatically and

31 Indeed, Kolodny makes a further point that is worth pondering: true immortality could make many of these things terrifyingly worse. Thus, he suggests, the redeeming feature of mortality might not be that it makes things *matter* to us but that it moderates *how much* the most terrible things matter (Scheffler et al. 2013, 166–167).

I'd need to visit them again. This applies to any other category of experience in a universe of constant change. The list of possible experiences is always growing faster than we can tick off the items it contains, which entails that even immortals must prioritize.

Kolodny makes a different, but complementary, point. Even in a lifetime with no ending, there will still be sources of temporal scarcity, since access to many opportunities and good things depends on timing in one way or another. I might need to prioritize obtaining a particular thing – or perhaps developing a relationship with a particular person – while I am at a particular stage of my own life (even if I will go through an open-ended number of such stages) or before someone else gets in first. Once again, even immortals – and hence the negligibly senescent – must prioritize, for which they will need to have some kind of hierarchy of settled preferences or values.

All of this analysis assumes that a modest conservatism is built into our present value systems. We don't want our own lives, or the lives of those who come after us, to be entirely without distinctive values, including the positive evaluation of virtues and relationships such as cooperation, friendship, moderation, fairness, and courage. Agar would, I think, see this modest kind of a conservatism as a cluster of species-relative human values, and in this case he might be correct. Be that as it may, I assume that it falls into *our* values to want to preserve a place for these virtues. Fortunately, we need go no further with this, since nothing suggests that negligibly senescent lives would be lacking in the way once feared by Nussbaum and more recently argued by Scheffler.

5. The great yawn of immortality

In his influential lecture, "The Makropulos Case: Reflections on the Tedium of Immortality", Bernard Williams argues that "immortality would be, where conceivable at all, intolerable" (Williams 1973, 82).

To set up the argument, he introduces the notion of "categorical desires" that "propel us into the future" (Williams 1973, 86): that is, there are things that we want unconditionally, or categorically, not merely subject to our survival. On this account, the badness of death is related to its thwarting of categorical desires, and for this reason, dying will typically be a misfortune for the person concerned. Eventually, however, these desires will come to an end even if life goes on. In response to this, Nussbaum has added that, in fact, there is *much* in human life – human "activities, projects, desires, hopes" – that death thwarts, making it unwelcome (Nussbaum 1994, 207). This seems right, but it is consistent with Williams' main line of argument. It seems possible that all of these things will eventually run out for us as *ennui* replaces them.

Williams refers to *The Makropulos Affair*, a play by the Czech author Karel Čapek, first published and produced in 1922 (and later adapted as an opera by Leoš Janáček). The main character, known by many names but always with the initials E. M., is, according to Williams, 342 years old, having been physically 42 for the past three centuries after using an alchemical concoction to prolong her life.[32] This extended her life by 300 years, but in the present she has grown cynical, emotionally cold, and unbearably bored. She is tired of living, and in the upshot she chooses mortality, allows another character to destroy the formula, and so denies herself any further period of life extension. Čapek's play suggests that greatly extended life would prove a curse for individuals and a source of disruption for human societies. For Williams, no desire or activity can be engrossing *forever*, so we'll eventually want to die. According to Williams, then, *The Makropulos Affair* shows how an immortal life would eventually be so dreadful that death would be welcome.

The only alternative, it seems, as we follow Williams' densely argued prose, is to live a series of psychologically disjoint lives while retaining the bare minimum of characteristics for staying the same person. After a brief analysis, Williams concludes that this is not something we could coherently wish for as a way of satisfying a wish to stay alive. At even briefer length, Williams dismisses another possibility: a series of varied lives in which a recognizable character is maintained throughout, or ultimately restored with an accumulation of varied memories. For Williams, this is an incoherent fantasy.

---

32 In fact, E. M. claims to have been born in 1585, to have taken the formula at 16, and to be 337 at the time of the play's action. See Čapek 1999, 238. She is apparently now physically 37, but for our purposes nothing depends on such details.

Despite expressing related concerns about whether we can retain our identies over very long lives, Shelly Kagan appears to favour the prospect of life extension within comprehensible limits. He suggests, however, that we could, in principle, reach a point where life has become cloying and oppressive, like eating chocolate beyond the point of satiation. There is, as he acknowledges, a further question of whether this in-principle possibility would ever actually happen if we were immortal, assuming that we retained youth, health, and adequate resources for our needs and desires. Thus, the question is whether or not to agree with Williams that life would always become unbearably boring if it went on long enough, perhaps for thousands, millions, or a billion years. Here, Kagan does agree with Williams, and he is quite persuasive when he goes into detail, including closely argued analysis of strategies to avoid eventual boredom.

These issues notwithstanding, Kagan emphasizes that it does *not* follow that it's good that we die when we actually do. It might still be that death "comes too soon for all of us" (Kagan 2013, 239), or at least for those of us who do not find ourselves in terrible situations such as from terminal and painful diseases. He concludes, "I don't believe I will have come close to scratching the surface of what I could enjoy by the time I die" and suggests that the best form of life would involve, not immortality, but "to be able to live as long as you *wanted*" (Kagan 2013, 246; Kagan's emphasis).

Note that this is an argument about true immortality. It's one thing to claim that true immortals would be bored after a billion years or so, if not before. It's another to claim that very long-lived people who enjoyed retarded or negligible senescence would succumb to boredom during much shorter lives in which they remain vulnerable to trauma, poison, and other causes of death that can strike down even youthful individuals. For this reason, Nicholas Agar is dismissive of the argument, even while agreeing that true immortality might be a curse. In particular, he observes, boredom from having done things too many times is the least of the worries of the very old:

> The oldest contemporary humans must endure the painful tedium of the aches and pains of aging, things that de Grey plans to abolish. They are more likely to experience frustration at an inability to do the things that they once loved rather than to feel bored at having done them too many times. (Agar 2010, 113)

Some philosophers have rejected Williams' argument even as it applies to true immortality. Fischer, for example, offers numerous examples of activities that he thinks are likely to retain their attraction indefinitely, within a reasonable balance of other things to do (Fischer 2020, chap. 7). In her 2013 contribution to the debate about immortality, Nussbaum concludes that E. M.'s plight in *The Makropulos Affair* seems more like an outcome of her own personality and situation than an inevitable fate for anyone who lives so long. Likewise, for Nussbaum, Williams' lecture seems like an expression of his temperament rather than a genuine argument. People with such temperaments might end up being miserable in a world of immortality, but that does not apply to all of us: "Nothing, however, has been shown about the livability of an immortal life. It looks better and better to me" (Nussbaum 2013, 41).

All that said, we are in territory where we don't really know the answer. It remains an empirical question whether or not we'd eventually find our extended lives boring, or whether radical longevity might bore some people but not others (Davis 2018, 41). Although he advocates research into radical life extension, Davis concedes that we might eventually be bored by almost everything that once interested us, which leads to the more precise question of whether or not "we can avoid boredom without changing our personalities and values over time and thereby avoid evolving into new people" (Davis 2018, 42). This prompts yet another question: are whichever pleasures might be repeatable, and might never pall no matter how long we live, sufficient to provide a full and satisfactory life? The fear is that they might not be, and that we might be forced to forget large parts of our lives to keep, say, studying philosophy fresh for us whenever we turn to it.

As one solution, Davis suggests a boredom pill – or more correctly, an anti-boredom pill – to make us feel engaged with whatever we're doing beyond the point where boredom sets in (Davis 2018, 43–47). Presumably, a society with the resources to provide radical life extension would also be able to invent and distribute such a pill. On this approach, according to Davis, your continued life might be objectively interesting and valuable, so there is nothing self-deluding, inauthentic, or otherwise morally questionable about taking a pill that makes it continue to *seem* so. This is certainly an ingenious argument, but it leaves me unconvinced. If I'm no longer propelled into the future by powerful desires, projects, and hopes, I see no point in taking a pill to *give* myself these things.

The fact remains that the argument from boredom has little force against radical longevity, even if it succeeds against true immortality. There is so much to do, if only we had the resources and energy, that I can't imagine getting bored in just a few hundred or a few thousand years. When our capacities flag, and ontological diminution sets in, yes, life might become burdensome, as we sometimes see with very old people who believe they've lived too long (I think here of my own father in his last years). But that is exactly what a technology of radical life extension would change.

6. Fear itself

Having dismissed the argument from boredom, Agar turns to an argument that we examined in a hypothetical way in Chapter Four – an argument about fear. He claims that someone with a negligibly senescent body would come to live a life dominated by the fear of death "to such an extent that it will prevent them from enjoying many of the activities that make our lives pleasurable and meaningful" (Agar 2010, 114). He paints a bleak picture of what this might be like, with negligibly senescent people developing aversions to travel, driving cars, sex, and general interaction with the world:

> Fatal car crashes are bad for senescing beings, but they're going to be so much worse for the negligibly senescent. A forty-year-old senescing human who gets into a car stands to lose, at most, a few healthy, youthful years and a slightly larger number of years with reduced quality. [...] At any time that a being who is negligibly senescent dies, an average of one thousand healthy, youthful years are lost. (Agar 2010, 116)

Agar observes that the negligibly senescent might enjoy surfing the Internet and indulging in activities, such as smoking, that are unsafe for us but would be safe for them. He also concedes that they would find other things to do and enjoy – things we cannot currently imagine. However, he argues that even this should not give us much comfort about the prospect of negligible senescence: "[W]e might adjust" (Agar 2010, 127), but we'd be adjusting to a mix of activities that we do not currently value.

If a very a long life were likely to be one constantly dominated by fear, that seems a powerful reason to avoid it. While de Grey might argue that a very long life dominated by fear is better than a shorter, more fearless life, that's a doubtful point. It seems to place too high a value on simply persisting, as opposed to enjoying the things that make life worthwhile: spontaneity, exhilaration and ecstasy, exercising our abilities to the full, the occasional feeling of being absorbed in the moment and free from care, and even an element of risk. As I argued in Chapter Four, the sort of extended life imagined by Agar might be one that we'd rationally reject, given our current value systems (though as we saw, the argument is inconclusive, and different people, with their own value systems, might rationally make different choices).

One problem with all of this, however, is that the argument depends upon a highly speculative empirical claim: that negligibly senescent people would have a heightened fear of death. It's noteworthy that Aubrey de Grey himself makes such a claim, and Jonathan Weiner has an amusing anecdote about this dating from their first meeting in 2002. As a passenger in Weiner's car, driving to Weiner's house on curving country roads in Pennsylvania, de Grey admitted to being frightened in cars: "That's how everyone will feel, he said, when they realize that they may have hundreds or even thousands of years ahead of them" (Weiner, 146). And yet, there is no good evidence that we do – or that we would if our lives were extended – fear death in proportion to the length of life that we might have ahead if nothing goes wrong.

In an article published in *Science* in 2006, Laura L. Carstensen summarized much of the research to that point relating to the effects on emotions and personality of a subjective sense of future time. One finding that she drew from the literature was that our sense of the fragility of life is primed by how far away death appears. For example, young men with acquired immune deficiency syndrome (AIDS) prior to the development of effective treatments had similar attitudes to the world around them to those of very old people who were running out of time. There was a similar effect when death became more salient as a result of dramatic events such as the September 11 terrorist attacks or the 2002–2003 SARS (severe acute respiratory syndrome) epidemic.[33]

According to socioemotional selectivity theory, "time horizons play a key role in motivation" (Carstensen 2006, 1913). Thus, our goals and prefer-

33 At some point, I expect to see similar studies of reactions to the coronavirus disease 2019 (COVID-19) pandemic that continues as I complete this book.

ences, and to some extent even our cognition, are affected by our time horizon as we subjectively understand it, and this operates as a psychological factor separate from life experience, the number of years already lived, and the physiological effects of maturing and senescence. If the remaining time seems short, we are likely to focus on short-term goals. If the remaining time seems open-ended, "goals that become most highly prioritized are most likely to be those that are preparatory, focused on gathering information, on experiencing novelty, and on expanding the breadth of knowledge" (Carstensen 2006, 1914). It seems, from the body of research reported by Carstensen, that older people are less adventurous than younger people, and perhaps less willing to take on challenges that have an unpleasant side.

In a searching analysis of Agar's argument, Aveek Bhattacharya and Robert Simpson respond that if he is right we should expect older people to become increasingly willing to take life-endangering risks, but in fact the opposite occurs. All the available evidence shows "that people are progressively less willing to take risks as they grow older" (Bhattacharya and Simpson 2014, 229). Furthermore, these authors cite converging lines of evidence that show how human beings tend to ignore very remote risks, including very remote risks of dying. These fall below a certain threshold of concern.

Given all this, it appears unlikely that negligibly senescent people would have a fear of death greater than that of senescing humans. Indeed, if death seemed to them to be more distant, its salience might diminish. From one point of view, we *Homo sapiens* are not entirely rational when we discount remote, but finite, risks of death. From another point of view, however, we seem to fear death more than can be rationally justified. Yes, death interrupts many projects, relationships, and other things we value, and that is undoubtedly a reason not to welcome it, and to want to go on living. But the deep fear or dread that most people feel when thinking of death goes beyond this, perhaps based on the seeming impossibility and enormity of the thought that the world will continue on its way while we cease to exist. If, as the evidence suggests, radical life extension would actually lower our fear of death, or at least our nagging awareness of it, that is a point in its favour.

7. Reasons (not) to be fearful

Many fears of radical life extension are exaggerated. It will not destroy structure and value in our lives, probably won't be boring (and might be the exact opposite), and won't make us paradoxically more afraid of death. That, at least, is the current state of the evidence. It remains true that this is a potentially radical technology whose full implications are unknown. If it extends human life by only a few decades, perhaps stretching the current 15 or 20 years of youthful adulthood to something considerably longer, and perhaps raising the longest lifespans to 150 or so, this is a relatively small, and seemingly benign, change – and objections to it might show status quo bias. Even this much change would force social adaptations (imagine a world where champion tennis players are still competitive at 65 or 70 and US presidential candidates are centenarians), but we can cope. The benefits described by John K. Davis and the reprieve from ontological diminution appear well worth the difficulties.

But the longer our lives become, growing to thousands of years or more, the more uncertain the implications. Nothing in this chapter is meant to suggest the contrary. It does not follow, therefore, that we need a Manhattan Project aimed at curing senescence. There's more to consider before drawing such a strong conclusion.

I've saved the Malthusian issues for a separate chapter. This is a Janus-faced topic, as it points in individual and social (even global) directions. If radical longevity greatly increases the world's population, that creates one set of issues, relating most obviously to environmental degradation and scarcity of resources. Attempts to avoid this could include draconian population policies that restrict the freedom to have children – and everything that goes with this in terms of family and intergenerational relationships. Here we have a more personal issue.

For Agar, this is one aspect of his claim that extended lives would be impoverished. What about the children? We'll tackle this head on.

Chapter Seven : What About The Children?

1. False consolation

Since the 1990s, Martha C. Nussbaum has become more sympathetic to the idea of immortality, no longer viewing it, as she did in *The Therapy of Desire* (1994), as inimical to anything resembling a human way of life. She continues, nonetheless, to look for consolation in the face of mortality. In her recent work, she suggests that there would be problems in any scenario for human immortality that we can imagine, whether it's a single individual becoming immortal, only a small group of immortals, or everybody gaining immortality. Although she advances her arguments in the context of true immortality, whatever force they have applies equally, or almost so, to radically extended human life.

In the first scenario, according to Nussbaum, the world's sole immortal would become deeply unhappy at "being unfairly singled out for an amazing benefit or having a piece of luck that is quite inappropriate" (Nussbaum 2013, 41) and experiencing "the sadness and isolation of outliving all one's loved ones" (2013, 42). In the second scenario, the result would be a horrible world in which the small group of immortals would oppress a larger underclass of mortals. In the third scenario, there would be Malthusian issues leading to draconian limits, or a ban, on having children:

> That world looks pretty bad too. It lacks all sorts of valuable activities connected with relations among the generations, and it also lacks a distinctive type of freedom to which we currently attach considerable importance. (Nussbaum 2013, 42)

Finally, Nussbaum adds an argument to the effect that if current adults accepted immortality they would be subverting the very system – with family relationships and other familiar aspects of human life to date – that brought them into existence. She claims that it would be inconsistent for someone to value the history that led to her becoming the person she is while also thinking "so lightly of that world that she wills it out of existence" (Nussbaum 2013, 43).

None of this seems consoling. After years of pondering the question deeply, and discussing it back and forth with John Martin Fischer, Nussbaum has come to perceive significant attractions in an immortal life that flows like an endless banquet with breaks for other activities and a certain share of troubles and dangers. She now views such a life as superior in many ways to ordinary mortal lives – but she thinks it is out of our reach. Although that seems plausible, it's no consolation to be told we can't have a better form of life because it would cause too many problems. All things considered, that *might* be a reason to abandon the quest for an immortal, or greatly extended, life. But we might feel disappointed, rather than consoled, if this were forced on us by circumstances.

2. Nussbaum's scenarios for extended life

To take Nussbaum's last point first, we subvert the past and existing systems that produced us every time we make social changes. The real question is whether those changes are for the better or at least neutral. In the case of radically extended lives, accompanied by a reduced role for family relationships, there is nothing inconsistent or hypocritical about knowing that your existence resulted from things as they were in the past, being glad that you were born, but also glimpsing what strikes you as a better way things could be in the future. Admittedly, it's a dramatic change in this case – but it might be dramatically *for the better.* That is open to doubt, but the previous chapter cast doubt on the doubters.

Still, this leads us back to Nussbaum's three scenarios. Is radical life extension ruled out in all three? In respect of the first scenario, I will grant that an individual person who becomes immortal – even in the watered-down sense of radical longevity – might ultimately find it a curse. The assumption is that this individual is one of a kind, and so is unable to find others who are similarly immune to senescence, the diminution of capacities and desires that goes with it, and the death it eventually brings. For a moment, put yourself in that person's situation. As everyone you care about ages and dies around you, perhaps not once but again and again with successive generations of friends and loved ones, you might be driven to despair.

I'm less impressed by Nussbaum's claim that this lone immortal would feel misery from an unjustly conferred benefit or "inappropriate" luck. To borrow from Nussbaum's own remarks about Bernard Williams and his argument from boredom, this sounds more like an expression of her personal temperament or anxieties than something I can imagine most people feeling. If extended life under these circumstances really were an amazing benefit, why not simply be glad to have it? Whether or not it is unjust, in one sense or another, could only be determined by settling questions of political philosophy that seem, for practical purposes, to be intractable. They are also, I suspect, uninteresting to most people who are not political philosophers. And what would be "appropriate" luck here? Saying that something is inappropriate assumes that there is an established standard of what is and is not suitable for a setting or an occasion, but we have no such standard for a case like this. Still, however much we might quibble, there's a large downside to the first scenario. It sounds deeply lonely and tragic.

The second and third scenarios are also unattractive – few of us would wish to see an elite of wealthy, unageing rulers lording it over a general population who live and die as before, and there would be a shock to the world's carrying capacity for humans if radical life extension were abruptly introduced for everyone. Overall, however, none of these scenarios is plausible as presented. A more realistic process of change would involve the gradual development of new life-extension therapies, adopted by more and more people over multiple generations. This would still be a radical change – a great global transition over a short time when viewed against the backdrop of human history – but it entails a smaller Malthusian impact than the sudden and universal availability of effective life extension therapies. Just how much smaller remains for consideration.

3. The shadow of Malthus

Although he is optimistic in many ways about life extension – and even true immortality – John Martin Fischer ultimately sees himself as a realist rather than an optimist, since he strongly doubts the likelihood of achieving "the environmental (and social) conditions required for" lives of indefinite duration (Fischer 2020, 138). He is especially troubled by the environmental challenges that we currently face, which threaten to make our planet a hostile zone for human beings:

> They have to be addressed soon; as I write this book, the glaciers and ice fields in Greenland and Antarctica are melting, and the seas are rising. Fires, storms, and floods are getting more frequent and worse. The world's nations have difficulty coordinating their behavior in the ways required to make genuine progress. (Fischer 2020, 138)

Fischer does reflect that if we were very long-lived we might be more responsible about environmental issues, since we could not ignore consequences likely to accrue only after our deaths. Nonetheless, he's inclined to rule out the right conditions eventuating any time soon for widespread extension of human lives. Similarly, Ingmar Persson and Julian Savulescu, who argue for research into *moral* enhancement[34] through technology, express concern about life-extension research:

> Such research tends to increase the acute population problems that already exist, and to enlarge the huge gap in life-expectancy that already obtains between people in the developed and the developing nations. (Persson and Savulescu 2012, 131)

Here, I am sympathetic. Widespread life extension now or in the next few decades would increase the world's population significantly beyond demographers' current projections of its peak later this century, further test the earth's carrying capacity for human life, worsen environmental degradation and global inequality, and make the challenge of anthropogenic global warming even more demoralizing, since each additional person on the planet will want and demand a Western lifestyle that includes plentiful access to electrical power and other forms of energy. For now, and for some time into the future, most energy will come from fossil fuels that produce greenhouse gases. Even now, our large-scale problems would be more tractable if the earth had fewer people. John K. Davis states plainly: "[W]e may already be facing a Malthusian crisis even without life extension" (Davis 2018, 105). The ge-

34 Moral enhancement is discussed in Chapter Nine below.

neticist David A. Sinclair, who has become one of the world's leading pro-longevitists, is similarly candid about the Malthusian impact of life extension technology (Sinclair and LaPlante 2019, 220–225).

Much has been written on environmental problems, and I won't add greatly to the mountains of paper. Suffice to say that these problems are serious, and an additional increase in global population – an increase not factored into current projections – can only make things worse. But the precise impact depends on timing, population policies, and human willingness to have fewer children than in the past. Unfortunately, we lack good predictive models of the impact of life-extension technology, and producing these would require estimates of numerous imponderables:

> This requires estimating when life extension will become widely available, how large the world's population will be at that point, what percentage of the human race will have access to life extension, how long people with life extension will live, and how many children they are likely to have. However, any estimates of these factors would be almost pure speculation. (Davis 2018, 105)

With the assistance of Shahin Davoudpour, a demographer who provided calculations of population trends for certain scenarios, Davis has analyzed some aspects of the issue in his 2018 book *New Methuselahs*. Taken at face value, this material confirms that widespread adoption of life-extension technology would have a significant impact on global population. The book helpfully publishes Davoudpour's calculations in an appendix (Davis 2018, 271–310), as well as explaining the methodology. With this material available for analysis and debate, there is now some basis to assess the Malthusian effects of life extension, which are often discussed in an abstract way, uninformed by any quantitative data. As Davis acknowledges, however, Davoudpour's calculations are highly simplified compared to anything that would happen in the real world, and they are offered in an effort to prescribe a set of numerical values (for birthrate, average lifespan, and so on) that looks sustainable when backed by a population policy that is crafted to produce them. While the calculations are too complicated to perform (or to check very far) without running a computer program, they still model very simple scenarios. We need to see calculations that include more factors and make more sophisticated assumptions.

Briefly, Davoudpour is not modelling the impact of radical life extension on an existing society that is in population equilibrium – its annual death rate matching its annual birth rate – or even shrinking (as in many current nations where fertility has fallen below replacement level). Rather, his figures show what happens if an entirely new population of babies is created and allowed to grow with no background death rate from people in earlier generations gradually dying. If this new population has a fertility rate at replacement level, it will eventually reach an equilibrium. The size of the population at equilibrium will not, however, be the same as that of the starting population. Its eventual size will depend on life expectancy and the spacing of births. For example, in the first scenario that Davoudpour analyzes, a starting population of one billion life-extended babies will stabilize at three billion after two or three centuries.[35]

The figures provided by Davis and Dhavoudpour are sobering, but they do not model what happens when life-extension technologies are gradually introduced into societies that are going concerns. Interestingly, however, even dramatic increases in an existing population might be manageable if the fertility rate can be kept at, or initially below, replacement level and the demographic impact of life extension is delayed rather than short term and abrupt.

The good news from current projections by demographers – without life extension factored in – is that the world's population will (probably) peak then begin to decline. This depends on whether current trends to lower fertility are sustained and whether they will drop to European levels throughout the rest of the world, including Africa. These are facts on which there is no firm consensus among experts. Using the figures provided by Davis, however, global population will decline significantly in the twenty-second century if the overall fertility rate stabilizes at the current level in Europe – about 1.5 children per woman in the population (Davis 2018, 113). If that should happen, global population will be about 3.5 billion by 2200 and about one billion by 2300.

35 This scenario assumes a 50:50 female–male split in the population, a life expectancy of 150 years, and two children born to each woman (one when she is 25 and one when she is 75).

Bearing in mind that the current population has passed 7.8 billion and is projected to peak at about 11 billion, this suggests that a widespread trend to radical longevity would be manageable if its effects were sufficiently delayed. Davis suggests that "it's not likely that life extension will be unavailable until 2200 – nearly two centuries from now" (Davis 2018, 113). But that remains to be seen. At this point, we have little idea how much progress is possible. Perhaps we can extend human life considerably with non-intrusive methods such as taking a cocktail of metformin and low-dose rapamycin, but perhaps not. Progress toward a life expectancy of anything like 150 years, much less something higher, is likely to be uneven, and some age-related causes of death might prove resistant to our efforts.

Moreover, a powerful technology of life extension could be available for some decades before its Malthusian impact is felt. First, it must be assessed as reliable, and then disseminated beyond a privileged sub-population with early access. Second, its demographic effects will become apparent only when early adopters actually start living beyond what would otherwise have been expected. For example, if a cohort of 60-year-old women who had an expectancy of another 25 years of life begins using a life-extending technology, this will not immediately upset demographic projections.

All of this suggests that a soft demographic landing is possible if powerful life-extension technology does not become universally available for another century or more. But this depends on many outcomes that we cannot predict with certainty, for example on whether we eventually see a strong downward shift in fertility in Africa, whether the current low rates of fertility in Europe and elsewhere are sustained, how human civilization fares over the coming century with its existing serious problems, and exactly how people will behave if their lives are extended in ways that grant extra decades – if not centuries or more – of youthfulness.

Even if we support radical life extension in principle, it probably cannot be disseminated widely in the near future – let's say during the remainder of this century and perhaps several decades beyond – without population policies across the world aimed at reducing fertility rates to extremely low levels. Depending on exactly what assumptions are made about timing, life expectancy, and other factors, these controls would probably be much harsher than China's widely condemned one child policy that operated from 1979 to 2016. Bluntly, such policies are untenable. There is no realistic prospect of any national population accepting them. They would need to be maintained for a significant time – of the order of a century or even a millennium, again depending on various assumptions – in the face of public resistance and subversion.

4. Life extension for the ten percent?

Davis supports urgent, large-scale funding to develop a workable technology of radical life extension as quickly as possible. He accepts that it would have a Malthusian impact, though he offers his own ideas about population policies that would mitigate this. He also believes, plausibly enough, that it would initially be used only by a relatively small sub-set of the world's population, probably the richest of the global rich. This leads him to some calculations of the positive and negative impacts if radical life extension were restricted to a small fraction – such as a tenth – of a society's, or the world's, population (2018, chap. 9). According to these calculations, if one tenth received enormous gains in years lived, this could outweigh any loss in quality of life for the rest.

Davis stipulates that the quality of life for 90 percent of the population is cut in half because of Malthusian and other impacts. He sees this as an assumption that is unfavourable to what he wants to demonstrate, so in that sense it is conservative. Let's accept it for now and assume that radical life-extension technology is introduced on a timetable that adds considerably to current global problems. Assume also that the "Haves" – the 10 percent who receive life extension – live on average for about 1000 years. Finally, we'll make a normative assumption, that a life of 850 years is ten times better than a life of 85 years of the same average level of welfare. That is, we assume (perhaps simplistically) that the goodness of a life is simply its duration multiplied by its average welfare over time.

While we can fiddle with various figures, my own (much rougher) calculations confirm that this works out just as Davis states. That is, I can't fault his arithmetic or change the result by introducing numerical adjustments that are clearly more plausible. The calculations seem correct as long as we retain the key assumption of a life expectancy of 1000 years for the Haves. That being so, the overall average goodness of lives within a population is improved if, as in Davis's scenario, a small minority benefits enormously from radical life extension while everyone else is left significantly worse off (even to the point where their welfare is slashed by 50 percent).

Davis can produce similar results if we choose to focus on the *total* goodness of lives in a population, rather than on the average. But that approach strikes me as obviously unintuitive. It can lead to repugnant conclusions in a well-known sense introduced into such analyses by the late Derek Parfit (Parfit 1984, chap. 17). Briefly: if we calculate totals rather than averages, a large population with lives of low welfare can be assessed as "better" than a smaller population with lives of high welfare. In what follows, therefore, I'll use averages across a population rather than worrying about totals.

What else needs to be assumed? Davis makes the normative assumption that we are seeking to produce the highest average goodness of lives within a population. As long as we employ a concept of the goodness of a life similar to that used in Davis's analysis – duration multiplied by average welfare – it follows that we are required to produce what otherwise seems an unjust outcome. That is, the fortunate 10 percent get enormously longer lives, and the other 90 percent end up much worse off than they would otherwise have been. Why does this seem unjust? Simply because those who are already best off, and seemingly least in need of getting even more, receive a great benefit, while the rest who are not so well off end up in a significantly worse position than when they started. Intuitively, that sounds outrageous.

Intuitions vary, and some readers might accept all the same normative and other assumptions as Davis. Others might import different assumptions about what makes for the goodness of a life, but these will be controversial. My response is rather different. I deny that we, as participants in a democratic political order, are under an obligation to produce outcomes such as this, where early introduction of truly radical life-extension technology greatly advantages some at the cost of harming many more. One reason we might reach that conclusion is that this is exactly the kind of intuitively unjust outcome that casts doubt on consequentialist moral theories. If the theory produces such outcomes, the thought might go, so much for the theory.

But here is another consideration that I give more weight. If the 10 percent are made enormously better off by a particular initiative, but at the expense of the 90 percent being significantly worse off, the interests of the two groups are no longer aligned (assuming they previously were, or appeared to be). Thus, the 10 percent have effectively become enemies of the 90 percent, who could reasonably deny any obligation to sacrifice themselves in the way that is asked of them. Such a scenario is a recipe for war between sub-populations, and we ought to avoid it at almost any cost. At a deeper level, the fundamental role of moral and legal systems and the apparatus of the state is *not* to maximize overall welfare, or the average goodness of lives, but to create a framework for social coordination and peace.[36] Within that framework, and subject to it, we might want to adopt policies in support of positive goals, such as a policy of researching moderate or even radical life extension, but this is more optional, and is subject to maintaining some commonality of interests within the population as a whole.

Note, however, that Davis and I both rely on controversial theories from moral philosophy. Again, for all I know, you might find his theoretical approach more plausible. If so, I have some last observations. First, Davis's sums assume that a life expectancy of 1000 years is achievable in exchange for halving the welfare of the 90 percent. But that cannot be known in advance. For all we know, only 150 years is achievable, and yet even this might have a large Malthusian impact, depending on other variables. That could change the calculations markedly. Furthermore, Davis's scenario assumes that the 90 percent component of the population will accept the situation they've been given (perhaps because they accept Davis's moral theory and have done their sums). In other words, they will peacefully accept a severe reduction of their quality of life. That is unrealistic, which means that there will probably be unexpected, unpredictable suffering for both groups.

In conclusion, I do not defend the abrupt introduction of radical life extension across a planet that is already experiencing a Malthusian crisis. If radical life extension is possible, we need to look for what I've called a soft landing. This is not as difficult as opponents make out, but it depends on numerous pieces of the puzzle falling into place.

---

36 This is a large claim, but I've argued for it in some detail elsewhere. See Blackford 2016, chap. 2.

5. The world of *if*

Radical life extension might work out smoothly, but a series of *ifs* stands in the way:

- *If* we delay creating any significant demographic impact (pushing it off until deep into the twenty-second century).
- *If* the world's population peaks and begins to decline, as most demographers expect.
- *If* human civilization survives global warming and other crises of our current century.
- *If* people with greatly extended youthfulness are prepared to reproduce at no more than replacement level (and perhaps initially at a level that is producing a declining world population).
- And of course, *if* our science is up to the task of inventing the required technology.

Each of these is challenging, but none is necessarily insurmountable. Climate change is a separate issue, but one that demands our thought and attention. The eventual success of science in retarding, halting, or reversing age-related decline remains to be seen. Relevant biomedical research will continue, no matter what, and it will make advances. At the same time, there are daunting scientific, logistical, and perhaps ethical issues in play that will affect the passage of time before life-extension technology could have significant effects on global population. Taking into account the Malthusian crisis that we already face, I don't support a Manhattan Project for curing senescence as quickly as possible, but valuable work should continue in gerontology without discouragement or stigma. Indeed, there may be room for increased funding of research into the fundamental mechanisms of ageing.

If, as a species and a global human society, we surmount these large *ifs*, our not-so-distant descendants may enjoy radical longevity. But this prompts questions, one of which Steve Fuller asks pointedly in his recent book on transhumanism, *Nietzschean Meditations*. Fuller suggests that the dramatic technological breakthroughs sought by thinkers such as Aubrey de Grey and Ray Kurzweil might not arrive in time to benefit those of us alive today. Rather, the beneficiaries might be our descendants after we're gone. He asks, therefore: "So, do you still believe in transhumanism even if it is unlikely that you will personally benefit from it?" (Fuller 2020, 202).

This might be a challenging question for many self-avowed transhumanists, but it is not one that I personally find troubling. In my case, I do not "believe in" any particular set of doctrines that might constitute transhumanism, or a version of it – that sounds too much like a religion for my taste – but I do hope and expect to benefit from improving biomedical technologies as they come on stream. Radical forms of enhancement, such as I consider in this book, will probably not be available in my own lifetime, though some modest successes in slowing senescence appear likely during the present century. For me, the question is about what choices, made now or in the short-term future, might be beneficial or harmful for human societies as they develop over the coming generations and perhaps evolve into a global posthuman civilization. I hope this attitude is widely shared, and that supporters of radical enhancement are not motivated solely by benefits to *them.*

6. But what about the children?

This leaves just one set of questions for the present chapter. Most centrally, how much discipline would people with radically extended lives exercise when it comes to having children? In *Humanity's End,* Nicholas Agar doubts that people who opted for radical longevity would adhere to a policy of remaining childless, but that is not necessarily the issue. If therapies for negligible senescence became an option in a world already under Malthusian stresses, one solution might, indeed, be a requirement that choosing extended life requires giving up on children, and I agree that this would not be acceptable to most people. Agar rightly states: "[Having children is] an almost universal aspect of human existence" (Agar 2010, 98). In fact this is an understatement in that the presence of children has been an aspect of experience in all human societies to date, and all societies have shown concern about the upbringing of children and their initiation into adult life.

If radical life extension does not begin to affect global population in any significant way until deep into the twenty-second century, however, it appears that no ban on having children will be needed. Depending on the precise timing and trends at the time, it might turn out that any population policies need not be terribly draconian. Nonetheless, it might be necessary to ensure that women would have, on average, no more than two children even if they had many years of fertility available.

This prompts its own scientific questions. Therapies that otherwise prolong youthfulness will not necessarily stretch the years of female fertility, since the latter depend on the availability of egg cells present in the ovaries at birth. Our understanding of female fertility centres around the concept that each baby girl is born with a finite number of ova that cannot be increased, implying that even very long-lived women would not be able to have babies beyond the current outer age limit (around the early fifties). One way to avoid this limitation involves collecting, freezing, and storing ova for later use, but it might also be possible to develop technology for women to retain or regain fertility in later years (Sinclair and Laplante 2019, 139–140). One way or another, therefore, radical longevity is likely to enable women to have children much later in their lives than experienced historically. In principle, there seems to be no upper limit to this.

In a high-technology society with average-length lives of 150 years, an average of 1.5 to two children per woman might be a plausible objective. But an average of only two children is harder to imagine if we're thinking about 1000-year lives with all that implies for numerous opportunities to reproduce. A world where people live so long while having children only at replacement level would look very different from ours. Imagine a medium-future civilization, about three hundred years from now. Radical longevity is almost universal, and some people are very old by early twenty-first-century standards while remaining youthful and strong. Global population is less than or comparable to today's, and the fertility rate is historically low even among the very long-lived. We don't see childless societies, but the spread of ages in the population is far wider than before, and widening all the time. One day, as everyone knows, it will range from birth to 1000 years or more. As time goes on, the proportion of children within the overall population is shrinking, and thanks to new kinds of parental choice people routinely have siblings decades older or younger than themselves.

Some readers might beg to differ, but this is not dystopian by the light of my value system. Children are still cherished, and intergenerational relationships continue, though in a different form from now, and as a smaller part of life. But whether it's dystopian or utopian, it's science fictional and looks rather alien. Then again, radical enhancement is radical, and no one has promised that everything else will stay the same once it appears in human societies.

In the past, a conscious desire to have children was not needed for the continuation of our species, even though many individuals have experienced and expressed exactly that desire. All that was needed was for most human beings to be fertile for a period of their lives, for humans to have an evolved liking for sex, and for most of us to attract mates. Until effective methods became available to separate sex and reproduction, babies were an inevitable by-product of fertility and the most usual forms of sexual pleasure. Now, however, effective contraception is freely available in most societies, and women in the most advanced economies have many opportunities in life that do not involve children. Fertility rates have declined dramatically across many regions, and large numbers of women and their sexual partners are opting out of having children altogether. For some women, this is a by-product of fast-paced modern life and career pressures during their fertile years, but for others it's a conscious decision to remain, as they say, child-free.

Having children – as opposed to nurturing and protecting whatever children we have – is being revealed in many societies as something less than a universal desire. It's a value held by many, presumably most, individuals in current liberal democracies. However, its intensity in an individual case and its frequency in a population depend on culture and socialization as well as biology. Nonetheless, it remains to be seen how realistic it is to expect a population of very long-lived and youthful people to reproduce at only replacement level. All of this warrants monitoring over the coming decades and centuries, but once we abandon the rush to benefit personally from radical life extension, possibilities open up for a soft demographic landing and a sustainable human or posthuman future. We can think more freely and speak frankly without the distractions of selfishness. This was the spirit of Godwin and Condorcet.

Chapter Eight : Threats to Identity

1. Leaving yourself behind

Life extension and cognitive enhancement sound good. Most of us would like to live a bit longer – preferably by stretching our years of youthful strength and vigour – and many of us would also like to be smarter. I know that I would, since I'm painfully aware that I have cognitive limits. I find some kinds of skills difficult to learn. Others fade too quickly when left unused. Across entire cognitive domains, I'm uncreative at best, or in many cases downright hopeless. In all those areas, I'd love some technological help.

This, however, sounds like moderate, not radical, enhancement. It remains roughly within historical human limits. So, what about something more extreme? What, for example, about living for centuries or gaining powers of intellect far in advance of known human geniuses? This sounds good, too, in the abstract, but it's harder to conceive of what it would be like.

Some smart, thoughtful people warn against these sorts of enhancement. You might, they say, gain years of life and plenty of extra brainpower, but in one way or another you'd lose yourself. This idea comes in several forms, but the main lines of argument point to the great differences between your unenhanced and enhanced selves. You'd change immensely over large stretches of time or via extreme upgrading of how you think. Either way, you might no longer be you, in which case the result is something like death. Even if you survive the experience, it might be as a person who is too different for you to identify with now, or even as someone you'd currently find repugnant – or it might mean living an alien and inferior kind of life as judged by your present values.

Interfering directly and radically with our cognitive powers sounds like tampering with our deepest selves, and we can scarcely imagine how we might change over vast spans of life. That being so, will we be the same people once we've gone through such experiences? In what follows, I examine some of the arguments against these forms of enhancement. In principle, I distinguish the arguments relating to life extension from those relating to cognitive enhancement. Some of the issues are, however, rather similar. Furthermore, life extension and cognitive enhancement might go together in practice. They could combine in a very long life with multiple cognitive upgrades. In fact, as we'll see, radical life extension *needs* an element of cognitive enhancement.

In all these case, say the critics, you should reject enhancement. You might amplify your powers but leave yourself behind.

2. Personal identity and radical life extension

The usual scenario that is discussed with respect to radical life extension tests how we should approach it if offered the appropriate therapies right now. This might reflect controversies over religious thinking about an afterlife, and hence the question of what religion can offer. Such debates have deep philosophical roots in Western and Eastern antiquity. In recent years, thinking about life-extension scenarios also reflects technologically grounded hopes of immortality, or the next best thing, for existing people who can afford it (Silicon Valley billionaires are often mentioned in this context). Recall that I counselled against this approach in the previous chapter. I don't expect radical longevity to be available soon, and I doubt that it will first appear in a form that completely halts senescence. Nonetheless, the arguments considered in this chapter relate to very long lives at any stage of a society. Are such lives inherently identity-destroying?

As with mind uploading – another contentious proposal favoured by many transhumanists – the personal advantages offered by radical life extension are an illusion unless they are advantages that you can look forward to enjoying.[37] That is, the key issue is whether *you* will continue to exist. But this takes us into notoriously difficult philosophical territory relating to the

37 For a summary of the issues, discussed in the context of uploading, see Blackford 2014b, 16–21. This is my introductory chapter to *Intelligence Unbound: The Future of Uploaded and Machine Minds*, co-edited with Damien Broderick.

question of how anybody can be the same person at different times of their life when they plainly do not have the same characteristics.

At least cases about living for a very long time do not involve baffling situations where individual people are imagined as fusing with each other, splitting in two like an amoeba, or somehow shifting from one physical instantiation (such as a human body) to another (such as advanced computer hardware). Nonetheless, ordinary cases of personal identity over time are difficult enough. How can I have been the same person at, say, the ages of 5 and 50? The answer to this question, if it were known and agreed, might tell us whether I could still be the same person at, say, 500. Derek Parfit (1984) suggested that the issue is not so much one of personal identity as survival, but this is almost as mysterious. In what sense did the 5-year-old me survive in the form of the 50-year-old me?

We might wonder just what is important to identity or survival in the relevant sense: what can be called survival as the same person. Is it bodily continuity? Is it a continuity of psychological characteristics, perhaps united in a certain way by chains of causation? Or is it what's called in the academic literature *connectedness* of psychological characteristics? The latter involves a commonality or similarity of characteristics (such as the same or similar memories) of the person at earlier and later times. Parfit suggests that what matters is a combination of psychological continuity and psychological connectedness, with little or no emphasis on bodily continuity – and with a heavy emphasis on the connectedness criterion (Parfit 1984, chap. 12). He also holds that survival (unlike the relationship of identity) comes in different degrees. On Parfit's theory, therefore, I might survive only to a small or negligible extent by the time I am 500, since I will have changed greatly and may have few memories and other psychological characteristics in common with my present self. It follows that I should not care very much about my 500-year-old self even if I'm confident that my body and mind will last that long in some altered form.

While this is only a quick sketch of the philosophical debate over personal identity and survival, it should be sufficient to show why critics of radical life extension find Parfit's ideas congenial. These ideas are very influential among philosophers, at least in the Anglosphere, though by no means forming a consensus view within the discipline of academic philosophy.

Ideas similar to Parfit's can be found, for example, in Shelly Kagan's view that we want more than survival over time as the same person. Kagan would not want to live for hundreds of years only to turn into someone with little memory of his current life and with very different psychological characteristics "even though I survive with the same evolving personality" (Kagan 2013, 167). Walter Glannon claims that it is not possible to survive as the same person for, say, 200 years. Too many memories would be lost, intentions changed, and the like, over such a period. Glannon also argues that, even if we relied on psychological *continuity* (rather than connectedness) as our criterion for personal identity, we could not be concerned about our greatly changed selves. Finally, he argues against the prospect of enhancing connections between psychological states such as memory, as these are in an equilibrium that "involves an optimal balance of mental states extending only so far into the past and only so far into the future" (Glannon 2002, 288).

Nicholas Agar distinguishes what he considers metaphysical and evaluative aspects of this argument:

> [O]ne can understand him as arguing that it's pointless for a human being to seek to extend her life span beyond 200 because facts about psychological connections show that no human being can live that long. Alternatively one can grant the possibility of extending human life spans to 200 years and far beyond, but insist that inspection of psychological connections show [sic] that this form of survival lacks much or all of what makes life worthwhile. (Agar 2014, 58)

Presented at an abstract level, none of this strikes me as persuasive. I doubt that psychological connectedness is as important as Parfit and others seem to think, and I doubt that bodily continuity is as *un*important as they seem to think. On the likelihood of changing over a long life, John Martin Fischer reminds us that we already change greatly during the decades of our unenhanced lives:

> I might now value challenge, change, and adventure travel, although I know that later in my life I will place higher value on security, comfort, and proximity to medical care. Whereas I am now progressive, I know that I may well later be more conservative politically. (Fischer 2020, 119)

An obvious comparison at this point is with growing up. People already change greatly during childhood and adolescence, but we don't usually think that a 5-year-old has ceased to exist, or failed to survive the experience of growing to adulthood, when she reaches later ages such as 18, or 28, or 82. Thus, Fischer identifies a double standard in evaluating the significance of change within mortal and immortal lives (recall that he is concerned with true immortality, but his argument translates to the context of radical longevity). He also identifies a specific double standard with memory. Even now, we might remember nothing about some earlier periods in our lives without being different people then and now, but this does not stop us from caring about ourselves as we'll be in the future:

> And yet our former selves can and do care about our future selves. I now care (especially) about my future self – the self that will (someday) retire and spend time with his family (and [maybe!] grandchildren), travel less, live within range of emergency medical care, and so forth. (Fischer 2020, 120)

When it comes to problems of identity, Fischer makes the point that he would be concerned about the possibility of being tortured at some time in the future, even after his values have changed (and likewise he would be glad at the prospect of winning the lottery at some time in the future, even after a change in his values). In each case, he states, "I care about this self because, well, it is me" (Fischer 2020, 119; compare Williams 1973, 46–63). This seems compelling. I grant, however, that there are no knock-down arguments in this area of philosophy and that intuitions vary.

At this stage, furthermore, I have only addressed the metaphysical aspect of Glannon's argument, as identified by Agar. I have not yet addressed what Agar calls the evaluative aspect of the argument, which appears more promising. I'll return to this, but let's have a first (for the purposes of this chapter) look at radical cognitive enhancement.

3. A magical and scary scenario

Threats to identity from radical life extension are long term in nature. Should you or I use this technology (contrary to my speculations about how long it will take to develop), our memories, our general values and motivations, and how we understand the world would not directly or immediately change. On the first day, nothing of that kind would seem different; rather, any issues arise over long stretches of time. At least in this respect, radical cognitive enhancement looks different from radical life extension.

We should keep in mind that radical cognitive enhancement does not mean enhancement within ordinary human limits, such as by taking a new drug that slightly improves our memories or sharpens our thinking. That would be a form of *moderate* cognitive enhancement, and it might well be beneficial. We can also imagine more significant forms of cognitive enhancement that still operate within, or largely within, human limits. By contrast, the idea of radical cognitive enhancement is transformation of our intellects to levels clearly and dramatically exceeding those of any human beings of the past and present (to the extent that we stand intellectually to ordinary adult human beings as they stand to non-human apes). We really have no idea of how this could be done, which makes it difficult to construct a plausible scenario. It's also uncertain how we would experience it.

Nonetheless, the debate in this area tends to assume a setting where existing, ordinary people in current societies go through a radical enhancement process. A different scenario would be a posthuman civilization at the end of a great historical transition. Here, everyone is born with cognitive abilities that are extraordinary by our standards, or perhaps they're equipped with upgrades soon after birth. I don't think there is anything wrong with such a society as far as it goes. But for the sake of argument, I'll assume for now that the issue is whether *we* should sign up for radical cognitive enhancement.

Perhaps this would involve implanting some kind of computer hardware, but even this might have limitations. Implanting a cybernetic device that imitates the functions of my cell phone, enabling me to perform a vast array of tasks just by thinking about them (not needing to use my fingers on a touchscreen keyboard), might have some advantages. But it leaves me much as I was when using my cell phone and my clumsy fingers. It offers, for example, no prospect that I'll devise a compelling Theory of Everything. If I want the genius of Einstein – and way beyond – something more seems needed, something that lies outside the current purview of science.

Thus, there's a sense, when it comes to radical enhancement of existing people under current conditions, that we're talking about fantastic scenarios and a magical technology. I feel this sense when reading Agar's discussion of radical cognitive enhancement in *Truly Human Enhancement*, although his discussion is careful and in many ways philosophically exemplary. When Agar writes of self-administering a form of radical cognitive enhancement that would, in the immediate aftermath, leave our memories and other psychological characteristics intact, this seems almost too much to entertain even in a discussion of speculative possibilities.

If such a technology were available, the most straightforward concern is that the process would immediately destroy anyone who went through it. That is, a different person would emerge. As Agar's description makes clear, however, such an interpretation is virtually ruled out. He writes: "[The technology] does not directly delete or modify autobiographical memories. It does not directly modify any significant beliefs or desires" (Agar 2014, 69). This sounds like a clear-cut case of survival as the same person, even if the process has what Agar portrays as dire longer-term effects. Immediately prior to undergoing the procedure, I'd have every reason to care about the person who would emerge afterwards because, to borrow Fischer's wording in the context of life extension, that person would still be me. If I somehow found out that I was going to be tortured after being enhanced, no one could comfort me by saying, "Don't worry, you won't survive the process. They'll be torturing a different person!"

According to this story, I might be startled or delighted by my new cognitive abilities, but I'd remember treasured moments (my first plane flight, my first visit to a foreign country, my first kiss when I was a teenager, or whatever those moments might be). My moral, political, and religious views would not immediately change. All seems to be well, at least until longer-term effects kick in. Yet, I would not trust this technology to work as advertised.

The trouble is that it promises to transform me in an extreme and drastic way in one quick process. I'm not sure how this would work, but I expect it would give me an enormously expanded knowledge base, dramatically improved and more fluid powers of reasoning, and a huge boost to the parts of my brain responsible for creative thinking. If all went well, this should quickly, if not immediately, give me astonishing and unpredictable insights into the world around me. With these, in turn, I might undergo a profound change in personality. My newly deepened powers of understanding might bring me great joy, but they might also, for all I know, bring severe depression or crushing metaphysical *angst.* This is not exactly like cyber-conversion or Borg assimilation, where an unattractive outcome is specified in advance. Rather, we *just don't know* what the psychological impact would be. It seems too much like a roll of the dice. Perhaps it might reassure me if there were already people who'd undergone the process – much as children can at least see something of what it means to become an adult – and had emerged from it without being traumatized and deeply troubled, but even then such a radical process might have different effects on different people.

Given all this, perhaps the most that I could meaningfully want from where I currently am, with my current values and understanding of the world, is some significant but *moderate* cognitive enhancement. If so, moderate cognitive enhancement might be repeatable over time, especially if combined with radical life extension. If I opted for only limited enhancement now, I might opt for further enhancement later, making that later judgement on the basis of my new experience and improving understanding. After various iterations of mild enhancement procedures, I might opt for a few years of experience with, say, the verbal facility of Shakespeare or the mathematical talent of Einstein. I might *then* move on to something more ambitious. In a futuristic world with radical life extension, I might end up with a superhuman intellect by installments. But right now, that is not something that I can comprehend or desire for myself.

Accordingly, I'll grant a claim at the outset to Agar and other critics of radical enhancement. It seems that truly radical cognitive enhancement, carried out in a quick process and in current circumstances where we have no experience of it or anything approaching it, is a genuinely frightening proposition. With that granted, I'll brazenly shift the goalposts, since moderate but significant cognitive enhancement is more realistic as a topic for discussion. In my partial defence, Agar's main argument seems to apply equally to this kind of enhancement.

4. Enhancement, memory, and transformative choice

To take stock of where we are, it seems likely that we would survive – as *ourselves!* – over the very long lives that would open up for us if scientists could cure senescence. Likewise, we'd apparently survive a process of radical cognitive enhancement, though the form in which we'd survive is unknown, suggesting that we'd be ill-advised to undergo a quickfire radical enhancement process in current circumstances. We'd want to know a lot more about its possible effects. Nonetheless, moderate, but significant, levels of cognitive enhancement have no such obvious downside.

Agar's critique of radical life extension and radical cognitive enhancement does not claim that either process would be immediately destructive of personal identity, and nor does Agar focus on the unknown immediate impact of upgrading our intellects to a truly superhuman or posthuman level. In large part, his concerns relate to the effects on memory, and hence what we'd be like after these procedures, setting aside the chance of any immediate psychological trauma.

When it comes to radical life extension, an important thread in Agar's argument is an empirical claim to the effect that the physical underpinnings of human memory are not adapted to lives of extreme length. This has some scientific support, and it seems to undermine the value of radical longevity beyond a certain (currently unidentified) point. Much as we have powerful reasons for attempting to delay or prevent senescence, those reasons would be weaker if the result was a life in which our brains could not cope with storing the memories of our potentially far greater sum of experiences.

Agar acknowledges that we could attempt to expand the brain's capacity to store memories, and to modify the processes by which memories are culled to make room for new ones. However, he adds, reasonably enough: "But this is a considerably more involved project than addressing the biological processes that are the proximal cause of aging" (Agar 2014, 59; compare Glannon 2002). Furthermore, it is well established that ageing brains show slower thought, poorer sensory awareness, weakening concentration, and faltering openness to novelty. None of this would be welcome to radically enhanced people.

Accordingly, research programs on radical life extension should incorporate research on certain kinds of cognitive enhancement. Organizations working on therapies to eliminate senescence should prioritize research on ageing memories and what I'll refer to for short as declining neural plasticity. Enhancing these will be a daunting task, and that alone might frustrate efforts to produce a technology of radical life extension in time for me, or probably you, to take advantage of it. In itself, this does not bother me, as I see no hurry to introduce radical life extension into current societies for reasons explained in Chapter Seven. I am not advocating the equivalent of a Manhattan Project to cure senescence, and I'm conscious of the Malthusian implications of radical life extension if it becomes available too soon. But the lesson at this point is simply that enthusiasts for radical life extension should regard appropriate forms of cognitive enhancement as very important: as important as slowing, halting, or reversing the body's more general processes of age-related decline.

Beyond his concern about the human brain's sheer physical limitations as a structure for storing memories, Agar fears that radical forms of enhancement could change us deeply in ways that we'd find unappealing or tragic if we knew about them in advance. This is partly because of a *different* set of effects on memory. These would apply in a radically extended life even if our memory capacities were expanded, and similar effects might be caused by moderate, but significant, forms of cognitive enhancement.

The worry about radical longevity relates to the greatly expanded time it provides for us to have new experiences, to change accordingly, and perhaps end up with different values, understandings of the world, and even personality traits. We might survive as the same people that we were, but much about our lives to date might be forgotten – not because we lack sufficient brain capacity, but because what we will come to regard as our early lives from the vantage point of 150 years, or 1000, or whatever it might be, will seem relatively unimportant. In the extreme, our memory of our first 50, or 500, years might end up being rather "thin" or sketchy. This is essentially Shelly Kagan's issue when he states that he wants more than survival over time as the same person. He wants to remain, in important ways, similar to his current self. We might, therefore, even with enhanced memory capacity, end up with rather different personalities and values from those we have now, and with memories that give relatively little importance to what has happened to us so far.

I don't find this especially frightening, and it's not like cyber-conversion or Borg assimilation, but in making choices about life-extension therapies we would want to be warned that this could happen. The alternative, of course, is accepting what happens now: a pitiable shrinking of our abilities and desires, leading to decrepitude, frailty, dependence, and death.

With cognitive enhancement, the concern is similar but not quite the same. The level of enhancement need not be, strictly speaking, radical. It's sufficient that it make us so much more cognitively capable as to look on our pre-enhancement selves from a position of clear superiority. For that, we needn't become super-intellects dwarfing the likes of Galileo and Einstein. It's enough to become much more capable than we used to be. Over time, we might come to feel disconnected from things that happened to us prior to enhancement. This might lead to us discarding some memories that we'd otherwise have retained (Agar 2014, 61).

If I were enhanced in this way, with this outcome, I would survive the experience and would not become a numerically different person from the person I was before. I would, however, develop differently after being enhanced than I otherwise *would* have. My new abilities would lead me to think, do, and experience things that otherwise would have been denied to me, and this would doubtless influence how my values and understanding of the world would evolve over time. I might come to have little regard for the experiences, ideas, and achievements of my pre-enhancement self. And so, I might even discard memories that now seem important.

The question is not whether I'd be unhappy with this situation after enhancement. I might well be glad that I made the choice I did, and I might not want to go back to how I was. The question to ask is whether, *right now*, I want to risk becoming such a person. This is a classic transformative choice.

5. Adulthood, transformation, and individuality

If I simply evolve over a very long time, with no direct technological tampering with my brain – except, perhaps, to help my memory storage and neural plasticity – this seems reasonably unproblematic. By contrast, significant cognitive enhancement appears to require a more invasive process, and it could lead swiftly to epistemic and personal transformations. Once cognitively enhanced to a significant extent, I'd be able, fairly much immediately, to regard my past life from a superior vantage point. That could lead to rapid changes in what I believe, value, and enjoy.

When making transformative choices, we cannot – and should not try to – step out of our current value systems. There is no genuine option of employing the values that we'll likely hold *after* making the choice. We must decide now, based on what we care about now. But for better or worse, our current value systems are complex, multi-leveled, and not entirely transparent to us. We can attempt to clarify our values, to think clearly, and to gather information, but to some extent we often end up taking leaps of faith when transformative choices confront us. This is unavoidable, given our cognitive limitations. Importantly, we value such things as being (somewhat) open to changes in our values themselves. We also value our continuing happiness after we make tranformative decisions. Taking account of this is not the same as applying the values we might have in the future. We care *now* about whether we'll be happy or miserable after transformative decision-making, whether we might feel fulfilled or empty, and so on.

In *Humanity's End* and *Truly Human Enhancement*, Agar compares and contrasts radical cognitive enhancement and the process of growing up. In *Humanity's End*, he refers to adulthood as "a stage that is supposed to mark the end of the dramatic intellectual and physical transformations that characterize youth" (Agar 2010, 185). I'm not sure what he intends here by the word "supposed".[38] Nonetheless, he's correct that human adults are generally more capable than they were as children. With some exceptions (learning new languages might be an example), they are better than children even at doing the things that children do, and like to do. The important difference is that most adults have no interest in such things as playing with bright plastic toys. In Agar's terminology, adults are "completely alienated" from those toys and the childlike pleasures they offer. Accordingly, "The intellectual enhancement that occurs as we mature certainly does not take away our ability to perform these acts; it takes away our desire to do so" (Agar 2010, 185).

The fear is that we'll be alienated from many aspects of our present lives once we judge them from a cognitively superior vantage point. What we *now* see as important aspects of our lives might *then* not even seem memorable. We might no longer find the same people attractive, fascinating, and delightful, or we might no longer find the same causes involving and urgent. If this outcome is something that we don't want, as it might well be, it counts

38 Supposed by whom? Agar and his readers? People in general? The government? God?

against getting enhanced. It needs to be set against the advantages that the enhancement process could bring, such as superior understanding and agency – which sound highly abstract but could open up all sorts of concrete possibilities for our lives. As always, we place a value on living with a certain openness to experience, and this can enable positive change even as it creates risks for changes in us that we'd now assess as negative.

It's reasonable to identify the risks, and hence a possible downside to significant cognitive enhancement, but Agar's argument goes further, and hence it overreaches. Agar suggests that cognitive enhancement to the extent of clear superiority would take away what is distinctive about us as individuals, since we would be likely to change our interests, attachments, and commitments in ways that we otherwise would not have. According to Agar, in *Humanity's End*, "We invoke a person's long-standing mature interests and commitments in explaining what defines her, what makes her distinctive" (Agar 2010, 186). In *Truly Human Enhancement*, he states that "adulthood is [...] a stage characterized by the physical, cognitive, and emotional resources to arrive at final and decisive plans about one's life" (Agar 2014, 72). Later, he expresses the point like this: "Something that we are right to dignify with the label 'life plan' – a considered statement of the best life for you and what really matters in life – arrives with human adulthood" (Agar 2014, 79).

But all of this is oversimplified, and it is largely false. Nothing so grandiose as a life plan typically arrives with human adulthood. Some adults may never be able to plan out their entire lives, whether because of their circumstances, their emotional inclinations, or the quality of their cognition. Others may possess the needed ability and find themselves in the right circumstances only much later in their lives. For a human adult, there may be further stages of transformation after growing up. We may develop new skills, gain knowledge (and learn entirely new ways to think), master a profession, change career (perhaps several times), become parents with all that it entails, move to another country or another continent, or undergo religious or ideological conversions (or deconversions). Furthermore, the effects of all this cannot be predicted. Even without enhancement technologies, then, human adults frequently make transformative choices that alter their attachments, interests, commitments, values, and worldviews – often dramatically.

Most sadly, perhaps, our connections with particular individuals can change for the worse. We can lose contact with people who were once important to us, or we can remain in contact but drift apart emotionally. More positively, however, we can also make new acquaintances who then become our friends, lovers, close allies, or collaborators.

Some transformative choices might change people for the worse, at least as assessed by our standards even if not by *theirs*. But we don't warn adults away from making transformative choices. Nor do we think these are, in themselves, tragic when they produce changes of values, attachments, and so on. Our individuality is not something static, though it can be difficult to define what *does* make us unique and distinctive individuals. Individuality can't, however, require that we stay the same once we grow up. This would rule out transformative experiences altogether for adult human beings.

What we might reasonably hope, on reaching adulthood, is not that we'll remain unchanged or take no risks with actions that might transform us. Rather, we might hope for changes that seem like enlightenment, growth, or liberation rather than corruption, harm, or decline – that was a lesson from Chapter Four. We might hope, too, that some deep-seated desires and traits of personality will change very slowly or hardly at all, and that these will influence how we respond to personal transformations such as learning to think like a mother, a lawyer, or a soldier. But this does not rule out swift changes, such as those from mastery of new concepts or conversion to new ways of understanding the world.

If we're transformed by direct technological interventions that enhance our cognitive powers, we might hope that this will not simply impose a new set of values on us, chosen from outside. Rather, we hope that our newfound abilities will interact with our existing psychological characteristics to produce *our* unique responses to being enhanced. To make this personal, I would hope not to become a standardized person – like a Cyberman – with new or upgraded abilities, but *me* with those abilities, adapting to them in my own way on the basis of psychological characteristics that I began with before the transformation. This hope is not unreasonable. Cognitive enhancement will increase certain of my abilities, but it will not alter me in a comprehensive way from outside to make me conform to someone else's template of what I should value and believe.

Significant cognitive enhancement is risky in that it might turn us into people we'd rather not be, perhaps losing some values that we'd rather retain. But many of us already take exactly those risks when making numerous other transformative decisions. Very often we do – and by our own lights should – choose the jungle of a new self rather than staying safely on the ship. Earlier in this chapter, I rejected the prospect of truly radical cognitive enhancement carried out through a quick process under current historical conditions. But up to a point, significant levels of cognitive enhancement might be attractive even after we've weighed the risks and the possible benefits.

In *Humanity's End*, Agar draws a comparison between radical enhancement and the effects of progressive dementia, such as experienced with Alzheimer's disease. If this is meant to apply to all levels of significant cognitive enhancement, it seriously overreaches. Agar observes that one thing which people fear about Alzheimer's is the loss of their intellects, and hence of their memories and reasoning ability. This is undoubtedly frightening, and although we might (for a time) survive extreme dementia its prospect evokes something of the terror of death. If I reached such a lamentable state, I'd be incapable of any of the activities that currently give my life a sense a meaning. Agar focuses on this point when he remarks: "But there's something else they fear as much and perhaps more [than the loss of intellect] – this is the severing of connections with the people and things that they value" (Agar 2010, 184).

An individual with advanced dementia may no longer recognize loved ones and may no longer understand social and moral issues where they had strong commitments. Such decline destroys everything that propels everyday, cognitively ordinary people forward in life, giving us something to look forward to tomorrow and the next day, and indefinitely thereafter. Significant levels of cognitive enhancement might put various of our attachments, values, and so on, at risk, but it is not fair to compare this to Alzheimer's disease. A better comparison might be with studying an intensive course, such as a law degree, that notoriously trains students in a new way of thinking: you learn, as they say, to "think like a lawyer".

Alzheimer's disease brings a radical diminishment of our powers, in which our ability to understand and act on the world is horribly narrowed. By contrast, cognitive enhancement increases our understanding and agency. Its risks might give us pause, and we might even, on balance, decide against it. But it is nothing like the death of personality and comprehension that is seen in the late stages of Alzheimer's.

6. Social transformative choices

Like individuals, societies change, sometimes in deep and pervasive ways. In Chapter Four, I offered just two examples of many that could be adduced from recent history. First, attitudes to war have changed in many, perhaps most, societies. We've largely renounced the warlike values of glory, military adventure, and conquest, though this was not something anybody planned. To many people in Western liberal democracies, it would have appeared an undesirable change not all that long ago. Second, our attitudes to sexuality have changed radically and are still changing. A vastly greater range of sexual expression is now tolerated and even welcomed, but again few people predicted this prior to the 1960s and fewer planned for it to happen.

The old values, however, were not universal human values, much as they might have been common in many societies. They were the values of particular societies in particular material circumstances, and when the circumstances changed the values changed. I'm sure that Agar would not classify the old attitudes to warfare and sexuality as "human" values or "species-relative" values that were worth retaining, but we need to be very careful with such concepts. The old attitudes could have been defended as human or species-relative values by their proponents, not many generations ago, who sensed that they were under threat. Such examples show the danger of specifying "universal human values" that ought to be respected and preserved.

Perhaps *some* values are universal across human societies, based on our evolved emotional tendencies. It's difficult to think of an example that is completely without exceptions, but we can generalize that human beings throughout history have valued the survival of their respective societies, the happiness of their particular loved ones, and perhaps certain aspects of their own flourishing. Any society's *mores* will have to accommodate these things. Some dispositions of character seem to be admired in all societies: for example, I doubt that any society values cowardice above courage.[39] Beyond this minimum, however, societies are likely to change greatly with changing cir-

39 Even here, however, what counts as cowardice in a particular society might depend on the society's general attitude to risk and/or on what specific forms of risk-taking behaviour by individuals are deemed necessary for the common good.

cumstances – moral systems can change, and so can legal and political institutions.

At any given time, some systems and institutions might be worth defending from ill-informed attempts to change or abandon them when cicumstances otherwise remain fairly constant. All the same, we should be open to changes in institutions and moral systems. Risks to these institutions and systems are not a good reason to frustrate wider change, though they might be a good reason to hope that change won't proceed too quickly for people to adapt psychologically and for societies to adjust institutionally. All of this involves matters of judgement, and reasonable people might disagree on what pace and scale of change is best. Agar is probably more open to social change than the bioconservative thinkers who served on the President's Council on Bioethics. I might be more open to some kinds of change than Agar, but I'm wary of the apocalyptic pace of transition desired by many radical transhumanists.

Elsewhere, I've defended technologies of genetic choice, including (moderate) genetic enhancement, observing that societies can absorb the impact of piecemeal technological innovation. Below a certain extreme pace and scale of change, tolerance of innovation "allows many kinds of people, with a wide range of values and interests, to flourish togther in the same society" (Blackford 2014a, 94). Compared to other innovations, including moderate forms of enhancement, radical enhancement has the potential for change that outruns the adaptive abilities of individuals and societies, and that is a reason for caution. But we mustn't become too defensive. Human beings have little ability to predict the future, and if our ancestors had possessed this ability, they might have been appalled by many changes that now appear benign. At the social level, we should, and often do, have a value of being open to change, even being open to changes in widespread values and the discourses that are used to embody and teach them.

Of course, there are some changes that we're *not* open to, and as we found at the end of Chapter Four there are no guarantees against them. None of us, I hope, wants to become a Cyberman or a Nazi, or to live under a fascist regime. These are not high-probability risks, but there's always some possibility, however remote, that massive cognitive upgrading or hundreds of years of life experience could somehow turn us to a darker path. Liberal democracies can be tempted to authoritarianism or worse, and no constitution provides a total shield. In unforeseen circumstances, individuals and societies can become obnoxious or malevolent. Anyone who wants absolute reassurance that this can be prevented will need to find a different book. Still, we can be vigilant, and remote risks like these should not stop us choosing good things.

7. Romance, social change, and the future

In *Humanity's End*, Agar discusses the risk to romantic attachments from radical cognitive enhancement, and indeed the chances of sustaining a romantic relationship would surely be lessened if one of the lovers underwent cognitive enhancement sufficient to leave the other far behind. Anyone contemplating this, while currently involved in a romantic relationship, might reasonably hesitate at the significant risk to the relationship. Agar adds that even a couple who go through cognitive enhancement processes at the same time are likely to diverge from each other as the range of things they can become interested in expands:

> The couple who enter the intellectual enhancement clinic hoping that their intellectual and physical compatibility and mutual love will survive the procedure is a bit like a couple who ask their respective employers to relocate them, and just hope that they get sent to the same city. (Agar 2010, 187)

But let's take this slowly. If we're talking about a magical technology that creates superhuman minds via a short, one-hit process, my best advice to the lovers is: "Don't do it!" But that's because such a technology sounds (to me) like a crazy idea and a threat to their peace of mind, if not their sanity. It's not as dangerous as being eaten by a shark, but don't try it at home.

If we're considering significant, but more moderate, cognitive enhancement, the situation looks more complicated. What if Jane and John, a pair of young lovers, emerge with increased brain storage for their memories – perhaps beyond previous human levels – and superior potential for mathematical creativity from targeted growth of neurons in certain regions of their brains? Does this create an unacceptable risk to their relationship? I'm not sure that it does. The experience might leave them more bonded than ever. What if, ten years later, they sign up for some more enhancement, this time raising their linguistic skills to Shakespeare-like levels? Again, I'm not sure that this is an unacceptable risk. Once again, their bond might even be strengthened.

Compare a situation from everyday life today. When Jane and John are 18 years old, they enrol for university study: Jane in law and economics, John in sociology, cultural studies, and comparative literature. Five years later, Jane joins a commercial law firm, where she negotiates contracts for large corporate clients. Meanwhile, John becomes a freelance journalist writing scathing reviews of computer games and Hollywood movies, published in left-of-centre independent magazines. For Jane and John, these choices are transformative. Their knowledge bases, skills, future opportunities, and values are likely to change dramatically. All of this might put their relationship at risk – in fact, they break up in their early twenties (years later, John will tell you that Jane sounds like "a right-wing shill"; let's not even start on what Jane thinks about him). Did they do anything wrong when they made these choices? Should someone have stopped them?

We already live in societies where there are many risks for romantic and domestic relationships, and it's not obvious that whatever forms of cognitive enhancement will become available in the near future will alter the overall risk level. That said, the social settings for romance and domesticity do change, as has happened in Europe over the past two to three centuries with the rise of romantic, companionate marriage as an ideal (Coontz 2005, chap. 9), and with more recent changes that have included easier divorce, the normalization of serial monogamy, the extension of marriage laws in many countries to include same-sex couples, and increasing, if still socially marginal, experiments with alternatives to monogamy.

It is likely that the settings will change further in any society of the medium or more distant future in which radical longevity is commonplace. In societies where people routinely live for hundreds or thousands of years, the ideal of lifelong monogamy might lose the considerable prestige that it retains today. It's unlikely that many couples would, could, or should even attempt to, remain monogamous over such long stretches of time, and this might be compounded if varied cognitive enhancement technologies make it easier for individuals to grow intellectually in diverse ways.

This might be a reason for romantic couples under current social conditions to hesitate to use certain technologies if they abruptly become available.

But it is not a reason to prevent societies from evolving more gradually, with new technologies influencing changes in social arrangements. It would be a mistake to view our current arrangements in Western liberal democracies as inviolable because they allegedly reflect universal human values. In fact, they reflect values that have a specific genealogy and arose in a particular historical, economic, and technological context. More generally, a wide range of practices and institutions that are accorded value in current liberal democracies might come under pressure as powerful enhancement technologies emerge over the coming decades and centuries. Exactly how this might happen defies attempts at prediction, but new or changed practices and institutions will emerge if given time.

Chapter Nine : Moral Status Enhancement

1. Moral enhancement of human beings

Since about 2008, bioethicists have given heightened attention to the possibility of moral enhancement of human beings in order to cope with contemporary global challenges (Douglas 2008; Persson and Savulescu 2008). The issue has provoked disputes even among academic commentators who are sympathetic to human enhancement more generally (see, for example, Fenton 2010; Harris 2011). Then, in their 2012 book *Unfit for the Future: The Need for Moral Enhancement*, Ingmar Persson and Julian Savulescu laid out their full case for moral enhancement, taking into account the controversy up to that point.

Following a common theme in the field of evolutionary psychology, Persson and Savulescu argue that *Homo sapiens* evolved in small, close-knit societies, and hence our psychological nature and core sense of morality are adapted for those conditions. However, they continue, we now live in societies with far-reaching power that extends across the world and into the future. Thus, we are faced with a mismatch between our evolved emotional tendencies (and the moral intuitions these give rise to) and the contemporary human situation: "We shall argue that human beings are not by nature equipped with a moral psychology that empowers them to cope with the moral problems that the new conditions of life create" (Persson and Savulescu 2012, 1).

For Persson and Savulescu, we face urgent global problems, and we might even perceive and acknowledge the urgency. Yet, we are not psychologically evolved to develop, embrace, and implement effective solutions. Our main problems fall into two groups: the proliferation of weapons of mass destruction (especially if they fall into the hands of terrorist organizations); and the phenomenon of global warming induced by human activities (along with environmental degradation more generally). These problems will require harsh solutions. In particular, Persson and Savulescu see a need for dramatic and permanent sacrifices of the privacy, individual liberty, and levels of material consumption currently enjoyed by people in industrially developed countries. But, they predict, these sacrifices will be rejected by citizens who value – and have become accustomed to – privacy, freedom, and material comfort. The only hope is if we can alter current human motivations, and as a consequence, adjust our current political ideals.

I don't, on this occasion, propose to examine the draconian policies that Persson and Savulescu advocate.[40] Nor will I consider the pros and cons of moral enhancement in the form advocated by these authors. Persson and Savulescu hope to overcome moral limitations that are, in their words, "so deeply entrenched in human nature that otherwise it might be hard to remove them" (Persson and Savulescu 2011, 443). Nonetheless, the sort of enhancement that they envisage would not involve any alterations to human capacities, including emotional capacities, that dramatically exceed the outer range of historically observed altruism, fairness, and other kinds of moral virtue. In that sense, these authors are arguing for a moderate form of enhancement. We will, however, need the concept of moral enhancement from time to time – especially to avoid any confusion between this and moral *status* enhancement.

Before we move on, it's worth observing that Persson and Savulescu have a broad understanding of moral enhancement. Any improvements (as they see them) to our motivations count as moral enhancement even if achieved through traditional, low-technology means such as education and appropriate institutional reforms. Thus, they do not insist on the use of direct technological intervention to bring about the enhancement that they envisage. They also acknowledge that it might not be possible to develop effective

40 As I write, the populations of many nations are enduring – and largely adhering to – severe restrictions on their freedoms, imposed by governments in an effort to slow the spread of the COVID-19 pandemic. However, it is too early to attempt any assessment of this experience or its significance for the Persson–Savulescu analysis. It would also be tangential to my purpose in this chapter.

technological options for moral enhancement during the time still remaining to tackle challenges such as global warming.

Despite their tone of urgency and their radical policy proposals, Persson and Savulescu defend a rather modest official position on moral enhancement research. It is simply that scientists should investigate the technological options. I don't propose to argue about that. By all means, I suggest, let's explore possibilities.

2. Post-persons and moral status

Nicholas Agar has little to say about moral enhancement in his critique of radical enhancement technology, perhaps because he sees it as a form of moderate rather than radical enhancement. He makes clear in *Truly Human Enhancement* that what he calls "moral disposition enhancement" is not his concern, and hence "I advance no claims about the feasibility or justice of disposition enhancements" (Agar 2014, 158).

However, more radical and exciting – and perhaps more frightening – possibilities appear when we consider the concept of *moral post-persons* (henceforth, just "post-persons"). These are beings who might come into existence as a result of radical cognitive enhancement or (perhaps) some kind of moral enhancement more drastic than Persson and Savulescu appear to have in mind. If a person were somehow transformed into a post-person, or if we could change the genetic makeup of embryos from a potential for personhood to a potential for post-personhood, these would be examples of *moral status enhancement*. So, what is a post-person? The idea is that post-persons are beings with a greater claim to solicitude and consideration – a "higher moral status" – than ordinary "persons" like us. They would, that is, have a greater entitlement to some kinds of beneficial treatment and to be spared some kinds of harmful treatment.

But we're none the wiser without a concept of what it is to be a *person*. Here, Agar offers a couple of standard definitions, borrowed respectively from John Locke and Immanuel Kant. The Lockean idea is that of "rational, self-conscious beings, who are aware that they have interests that persist over time", while the Kantian idea is basically that of beings "with a capacity for practical rationality" (Agar 2014, 159). Although there are other approaches,

I'll stick with these two, since they introduce more than enough complexity for this discussion. We already have in play four difficult concepts: rationality, self-consciousness, awareness of oneself in time, and Kantian practical reason (or practical rationality), which is not necessarily the same thing as rationality in its ordinary or general sense. We might hope that all of these would be found together, perhaps emerging in creatures with a certain level of cognitive ability, but that has not been demonstrated.

An approach to personhood that Agar does not mention – and I bring it up now only to set it aside – is contractualist in nature, relating to the ability of persons to accept, and to act in accordance with, relations of mutual accountability. This relates to a Kantian conception of practical reason or rationality, but the worry, once again, is that these various characteristics do not necessarily translate into each other or go together. There might be highly rational beings, in the sense of possessing extraordinary problem-solving abilities, that have little or no self-consciousness and no mutual recognition of each other of the kind favoured by contractualists. Other such combinations might suggest themselves to readers. My emphasis in the remainder of this chapter will be on self-consciousness combined with awareness of oneself as having a future that can be planned. Intuitively, these do seem to fit in the same package. In any event, I assume for the sake of argument that personhood is something of a package deal.

Many philosophers have thought that persons are more worthy or deserving of moral consideration than creatures which are merely sentient (able to feel such things as pleasant and painful sensations). Again, many philosophers have thought that sentient creatures are more worthy of moral consideration, all else being equal, than creatures (whichever they might be) and other things that lack the capacity for sensation. These concepts are sufficient to equip us with a classic three-level theory of moral status, dividing the world into the following categories: insentient non-living and living things incapable of experiencing pleasant or painful sensations (these are said in some accounts to have *no moral status*); living things capable of experiencing pleasant and painful sensations (perhaps with special importance attaching to their vulnerability to physical suffering) but not qualifying for the third and highest category (these are sometimes said to have *some moral status*); and finally, living things that, in addition to being sentient, possess certain sophisticated psychological characteristics such as those identified by Locke and Kant (these beings are sometimes said to have *full moral status*).

A post-person, then, would belong to a fourth, higher, level. A post-person is not a person whom we'd regard as unusually virtuous, such as an individual with dispositions of character that we admire and look for in our acquaintances and fellow citizens. It is not even someone who's been brought to a level of virtue that we'd regard as *markedly* superior to our own via one or another kind of moral enhancement technology. Rather, a post-person would be someone or something whose interests deserve more consideration than those of "mere" persons. Agar seeks to argue that post-persons could exist, but he concedes that it is very difficult for us to imagine what they would be like. He therefore considers an argument from Allen Buchanan to the effect that we have no clear conception of moral status enhancement.

According to Buchanan's argument, as Agar formulates it, "A natural explanation for this inexpressibility is that there are no such statuses [higher than personhood] and nor could there be" (Agar 2014, 158). Buchanan's point is that "the concept of a moral status is a threshold concept" allowing for a wide range of inequalities, "including inequalities in the very characteristics that confer moral status" (Buchanan 2009, 359). Thus, we would not normally think that *enhanced* levels of understanding of ourselves as agents operating in time, with a past and a future, would turn us into beings with a new and higher moral status. On the classic theory, you either possess the relevant characteristic or you don't. Once you do, you've crossed the threshold to personhood, and we know of no higher moral status than that.

Agar concedes the existence of moral statuses, each with its own threshold, since, he thinks, the obvious alternative would give more support to his concerns about radical cognitive enhancement. Thus, he argues from what he regards as premises that are favourable to his opponents in this debate. The alternative is what Buchanan calls an "interest-based view" (Buchanan 2009, 360) with no moral statuses in the strict sense. Instead of three levels, as in the classic theory, there is a gradient, or a sliding scale, from beings with less to those with more entitlement to moral consideration. This gradient "ranks beings according to the different amounts or degrees of good their lives involve" (Agar 2014, 160). For example, on the interest-based view, human beings have a stronger interest in surviving than a dog because of our superior rationality and imagination, and radically enhanced individuals might have a greater interest in surviving than *us:*

> Presumably the being with powers of reason and imagination far superior to those of humans anticipates death in a more vivid way than we do and therefore has a stronger interest in avoiding death. Harms that fall short of death derail post-persons' projects that are more complex than any of those that we undertake. (Agar 2014, 161)

If we adopted the interest-based approach to who or what deserves moral consideration, we could easily conceive of beings who possess far *greater* capacities than ours for such things as understanding themselves as existing in time and having a future ahead. If these beings adopted the interest-based approach, they might consider themselves more deserving than we are of moral consideration. By contrast, the classic theory of moral status assigns us and these beings to the same category. We are all persons, and we are all entitled to the same level of consideration. To stake a claim to a higher level of consideration, the individuals with greater capacities than ours would need to point to a qualitatively new capacity.

For the moment, let's proceed on the basis that the classic theory is correct as far as it goes[41] and that personhood involves a package of related characteristics. On these, admittedly debatable, assumptions, is it possible to imagine a further moral status that is higher than personhood?

3. Can there be post-persons?

To say the least, it is not clear what interests or abilities would define a fourth level that should be added to the classic theory. Accordingly, Buchanan concludes that, the more we accept the existence of moral status in the strict sense, "the more difficult it is to imagine how biomedical enhancements could produce beings with a higher moral status than that of persons" (Buchanan 2009, 368).

---

41 Only "as far as it goes", since the existence of post-persons/supra-persons would transform the three-level theory into a four-level theory.

In response, Agar acknowledges our inability to imagine a new moral status and, with it, the nature of post-persons, perhaps because of our limited cognitive powers. He nonetheless develops an argument that a higher moral status could exist. At the same time, he criticizes various published conjectures as to what characteristics might support this higher status. For example, Jeff McMahan has raised the possibility of "supra-persons" (McMahan 2009, 600) whose higher moral status is based on a form of cognition greatly superior to that of human beings. At some point, on this approach, superior cognition crosses a new threshold. So, what would supra-persons be like?

McMahan mentions two possibilities: perhaps they would possess libertarian free will even though, as it happens, *we* do not (McMahan 2009, 603–604); or perhaps they'd possess a kind of super-empathy that we lack (2009, 604), which sounds like an extreme form of *moral* enhancement. It would involve actually experiencing, and then being able to reflect upon, the mental states of others.

The free will idea faces metaphysical objections. As is well-known to aficionados of the relevant debates, the idea of libertarian free will is so mysterious that we might wonder, as indeed McMahan does, whether it is even coherent. If it exists, libertarian free will involves some kind of ability to make choices that are neither causally determined by the state of the world (which includes our own physical composition and psychological make-up) nor the outcomes of pure chance. If we have libertarian free will, at least some of our choices are truly *ours* without flowing from what we are actually *like* at the relevant time and in the relevant circumstances (see, for example, Strawson 2010, 22–26). If all this is coherent and scientifically possible, it's baffling why we don't already possess libertarian free will although cognitively superior beings might. As an illustration of one conceivable kind of higher moral status, the proposal is ingenious, but it is flatly unconvincing.

By contrast, I am unsure how to regard McMahan's idea of a kind of super-empathy. This is at least intriguing, and seems less mysterious than libertarian free will. It encounters the problem, however, that it does not necessarily create a new vulnerability in the supra-persons, or alleged post-persons, that McMahan describes. Recall that sentient beings are vulnerable to physical pain, while persons are vulnerable to pain and to a wide range of additional kinds of harm and suffering – for example, feelings of grief and deep sadness when their hopes about the future are ruined. These seem to be vulnerabilities that are worth taking into account when we deal with sentient beings and with persons. Nothing like this – a qualitatively different kind of vulnerability – seems to arise from the capacity for super-empathy. The example, therefore, does not ultimately convince me, but it's another clever suggestion, and it should at least encourage us not to *rule out* the possibility of a higher moral status than those identified in the classic theory. All the same, I continue to find the idea of an additional moral status unpersuasive, even assuming the truth (as far as it goes), of the classic three-level theory.

Agar is also unconvinced by McMahan's examples (Agar 2014, 171–173), and he acknowledges that we lack an adequate understanding of what beings with a new and higher moral status might be like. Nonetheless, he argues that a higher moral status – or a cascade of such statuses – would emerge from sufficiently powerful cognitive enhancement. One aspect of this argument is Agar's claim that cognitively enhanced beings could have powers of practical moral reasoning that lie beyond our own and need have no upper limit. While there is a limit – however remote it might be in practice – to how much can be known about a particular language, this is not so with practical moral reasoning: "It differs from knowledge of English in making no indispensable reference to principles or ideas known by a community. It's something that can always be improved" (Agar 2014, 178). In this respect, he suggests, it is more like knowledge of mathematics than knowledge of a language. Somewhere in the endless expanse of possible improvement to practical moral reasoning, we should predict a threshold to a higher moral status (Agar 2014, 179).

Considering the range of texts that have been created in English by such masters of the language as William Shakespeare and James Joyce – among many others – this strikes me as an odd way to characterize knowledge of English, or of any language that has had similarly great virtuosos. Still, it's possible, for all I can tell, that Agar has identified a genuine difference between knowledge of a language and knowledge of mathematics. Perhaps languages are bounded, even if their boundaries are very wide indeed, in a way that mathematics is not. But even if that's true of mathematics, I doubt that there's a good analogy with practical moral reasoning.

For a start, knowledge of the complexities of a particular culture's moral system might be something like knowledge of a language, and formal investigation of an existing moral system might require similar skills to those of a grammarian or a lexicographer, notwithstanding that the practical application of the system might allow for as much creativity as the use of a language. In fairness to Agar, however, he is not thinking of skill in mastering a particular society's moral system. He might be assuming the existence of a single true, or objectively authoritative, morality, or more likely, given his broader philosophical commitments, a single and universal *human* morality. But even if there is such a thing as a single true morality in one of these senses, which I doubt, the analogy with mathematics is unconvincing. Perhaps the true morality – in whichever sense Agar prefers – is a finite system of rules and principles that we have not yet fully described, but that could be known, in principle, in a similar way to a language. Thus, applying either an objectively authoritative morality or a universal human morality to the multitude of practical situations that arise in human lives would be a finite, though creative and difficult, task.

In all, I am unconvinced by arguments for the possibility of post-persons. Agar acknowledges "that we won't understand exactly how enhanced cognitive powers would grant a higher moral status" (Agar 2014, 179), and this is tantamount to conceding that we cannot think of a truly convincing illustration of what post-persons would be like. He seems to be swayed by a hunch that the existence of three levels suggests the possibility of others, but why *shouldn't* there be exactly three levels in our best theory of moral status? If we could identify four levels, should we thereupon infer that there must be a fifth? Once we start inferring that there's likely to be another level, because we already know of three, we seem committed to believing that there is (probably) no end to the number of levels. Agar appears to accept this implication, but the result would be something very different from the classical theory of moral status.

We know what it is to suffer pain or experience pleasure – as merely sentient creatures can do. We also know what it is to anticipate the future, as most human beings can do, which entails plans, projects, worries, frustrations, and fears about what the future might bring (in contrast to the raw feeling of fear when placed in immediate danger), and we understand the possibility of deep sadness or grief when our hopes for the future go wrong. We can imagine all this applying to a greater degree to cognitively superior beings, but we have no inkling of any further set of interests that could exist that is *different in kind* from those of sentient creatures and persons. Accordingly, we can find some support for the idea that the best theory of moral status requires three levels and no more: no moral status; some moral status; full moral status.

4. Problems for the classic theory

One problem for the three-level theory is that the interest-based view, or perhaps a modification of it, has obvious strengths even though Buchanan finds it unattractive and Agar sets it aside. Unless we embrace recondite ideas of panpsychism and the like, most entities, including many living things, can be presumed to possess no consciousness at all. But we might think that there is then a sliding scale of consciousness and sentience, which gradually blurs into more sophisticated kinds of self-consciousness and self-awareness, including increasingly sophisticated ability to conceptualize and imagine the future, make elaborate plans, and feel the sorrow that accompanies lost opportunities. On any such account, different beings can have weightier or less weighty interests because of their particular characteristics. Agar suggests that this sort of approach could favour people with radical cognitive enhancement over ordinary humans – for similar reasons to its preference for humans over dogs. Radically enhanced people "with powers of reason and imagination far superior to those of humans" could anticipate death more vividly than we do, and would have more complex projects than ours for death to cut short (Agar 2014, 161).

Even if we can't meaningfully represent all this on a single scale where, in principle, we could find a location for every entity in the universe, it suggests that there are plausible ways of thinking about moral status – or at least about what kinds of things and beings merit what kinds of consideration – in addition to the classic three-level theory. However, this is not the deepest problem for that theory. More fundamental is the idea that moral statuses are *assigned* by human moral systems rather than found in nature. Let me explain, as this is conceptually tricky.[42]

42 For a more elaborate critique of the whole idea of moral status, based on somewhat different considerations, see Horta 2017.

It's true that some beings possess capacities and vulnerabilities that others do not, and so we are responding to something real when we treat plainly sentient organisms (which at least includes all mammals) with more regard, or a different kind of regard, than we give to a life form that we know, believe, or hope to be incapable of suffering. We would be most reluctant to damage a magnificent and irreplaceable plant, such as California's Hyperion redwood and General Sherman sequoia, or Mexico's Tule tree, but this is not because we think that the plant can suffer in any way. Rather, we treasure it for its aesthetic value to *us* and potentially to posterity. We might even think, correctly or not, that it possesses a sort of value that should be respected by any rational being. Either way, our solicitude for the welfare of creatures that have nervous systems similar to ours – and are hence vulnerable to similar experiences of pain – is qualitatively different from how we think about plants. So far, this supports the classic three-level theory.

Likewise, we are responding to something real when we show a different kind of solicitude to creatures that seem capable of deep sadness and grief, including grief caused by dashed ambitions, precious but crushed relationships, and the like. As far as we know, only human beings – and only human beings who have grown beyond infancy to an extent that has never been defined precisely – possess this kind of vulnerability to a significant degree. On the other hand, it is likely that some other mammals have the beginnings of it. Among these creatures are apes and cetaceans, possibly elephants, and perhaps also domestic dogs, which have been bred for millennia as companions to human beings. Thus, we show a form of solicitude to human children and adults, and to a limited extent to certain non-human animals, that is qualitatively different from the constrained solicitude we give to creatures that are vulnerable only to physical pain. Once again, we are responding to objective features of the natural world, even if we are unsure which creatures – and even which infants and young children – can have the relevant experiences and which cannot.

To some extent the observations in the previous paragraph support the classic three-level theory, but the differing levels of self-consciousness that we suspect among intelligent mammals also cast doubt on it. It's not as if we assign the same moral status to all creatures that might, even to a rudimentary extent, be self-conscious and aware of themselves in time. Perhaps we should, but this is not reflected in our actual moral practices.

Importantly, we also assign a high moral status to newborn babies, even though they lack sufficient self-awareness and psychological sophistication to count as persons. Babies seem to experience an urgent kind of frustration when their immediate needs and demands are not met, but they are incapable of anything as sophisticated as feelings of grief, which require higher levels of understanding than they can muster. It appears that our emotional inclinations and moral systems compel us to show great consideration toward very young children because of what they mean to us (among other things, they represent the future of our families and our societies) and out of sheer ingrained responsiveness to them. Put plainly, most of us find babies heartmeltingly cute, and we cannot imagine anyone wanting to hurt them. This overrides the empirical fact that they are *not*, based on Lockean or Kantian criteria, little persons.

These considerations are consistent with a point that is now accepted by many philosophers and other scholars. Human moral systems do not map a reality that transcends human desires and institutions. Human moral systems do not even need to be completely internally consistent, as long as they're consistent enough for their function. They are, in brief summary, a form of social technology that assists social coordination, social survival, and internal peace (see Blackford 2016, chap. 2). In an article published some years ago now (Blackford 2007), I argued that the defining characteristics of Lockean personhood, in particular, are real and important. They track significant differences in the vulnerabilities of creatures that do and do not possess these characteristics. I'd now add, however, that vulnerability to pain, and to emotions such as grief, is important to *us*, not to the mind-independent universe. Other things are equally important to us, or more so, not least the survival and success of our own families and societies. It should not be surprising, therefore, when human beings show more concern for the welfare of a newborn human baby than for that of an adult gorilla which might actually have stronger claims to be a person in the Lockean sense.

I conclude that we unenhanced *Homo sapiens* are unlikely to assign a higher moral status to aliens or posthumans – or to any other intelligent beings that potentially compete with us – than we assign to ourselves. Even if some group of posthumans were vulnerable to a form of suffering beyond pain and grief, and hence beyond our experience and comprehension, we would find it impossible to identify with their suffering. Furthermore, if their interests were hostile to ours we would give our own interests priority.

In summary, the classic three-level theory of moral status tracks characteristics that are objectively real, without including *all* the objectively real characteristics that shape actual moral systems. The levels of moral responsiveness described by the theory, based on particular characteristics such as sentience and self-consciousness, are not objectively demanded of us, and in practice they can be outweighed by other features of a situation, such as what conduct is likely to preserve ourselves, our offspring, or our society. To conclude this stage of the argument, there are powerful reasons to think that *we*, as unenhanced (or if it comes to that, moderately enhanced) human beings, are unlikely to recognize any group of posthumans as post-persons who are therefore entitled to greater consideration than we give each other. Nor are we rationally compelled to do so.

5. Would posthumans *think* they are post-persons?

Posthumans might still be *Homo sapiens* according to the biological species concept, but if they're sufficiently enhanced they might come think of themselves as forming a separate and superior population. If they're mutually bonded and clearly recognizable as a breed apart, they might not view themselves as just another group of persons, any more than we humans view ourselves as just another group of persons in addition to other (possible) groups such as apes and cetaceans.

These posthumans could hardly deny our capacity for pain and grief. They would have the same evidence that we do that many living things, in addition to themselves, and importantly including us, are sentient. Furthermore, they would have overwhelming evidence that unenhanced humans like us (and any moderately enhanced humans who live among us) can feel intense grief, even to the extent of taking our own lives or engaging in other self-destructive conduct when things go badly wrong and the future appears hopeless. The posthumans' acquaintance with human conduct and the products of human cultures would give them ample evidence of our kind of consciousness, and we might hope that they'd try not to cause us pain or grief. Since their psychological make-up would (I assume) descend from ours, this might be a reasonable hope. Nonetheless, they might regard themselves as post-persons or employ some comparable concept.

As usual, scenarios matter, and Buchanan usefully distinguishes the two obvious scenarios relating to post-persons. In one, radically enhanced beings who've become post-persons have the rights appropriate to them in a society of such beings, and there is no problem. Buchanan sets this to one side, apparently regarding it as uninteresting, but it provides a reminder that many of the problems relating to radical enhancement arise within what could turn out to be transitional scenarios. As always, we need to distinguish between problems in *transitioning* to societies made up of radically enhanced people and problems in futuristic societies *of* radically enhanced people.

In the other scenario that Buchanan describes, "some human beings are enhanced to the point of becoming post-persons, but others are not" (Buchanan 2009, 349). Here, the problem is that some beings with the characteristics required for personhood will have greater rights than others, since some will have the superior rights (whatever they are) appropriate to post-persons. For the remainder of this chapter, I will focus on scenarios where actual or self-proclaimed post-persons coexist in the same society with humans who might even be enhanced to some degree but are clearly not post-persons. As we'll see, problems arise even if some cognitively advanced posthumans merely *regard* themselves as post-persons, and perhaps even that much is not required for some undesirable outcomes.

At one point in *Truly Human Enhancement*, Agar stipulates what he describes as an orderly transition to a mixed society of persons and post-persons, by which he essentially means that there are no, or relatively few, violations of anyone's moral entitlements, based on their status as persons or post-persons: no "high levels of directly treating citizens in ways incompatible with their moral status" (Agar 2014, 182). Agar's argument is that, even within this relatively benign setting, the interests of post-persons would rightly trump those of mere persons whenever there was a clash. Post-persons would, we might suppose, identify many situations as warranting the sacrifice of ordinary persons. Furthermore, on the assumptions that Agar is making, they would be correct. On those assumptions, we should "exempt from moral condemnation the particular decisions of post-persons to inflict harm" on mere persons (Agar 2014, 191).

Agar's argument, therefore, is not that the post-persons in this mixed society would act wrongly but that problems would result from upstream actions that Agar definitely does regard as morally wrong, namely the actions that allowed such a society to come into existence in the first place. On Agar's approach, *those* actions should be condemned because they would (probably, or perhaps inevitably) produce bad outcomes for mere persons. Since there is no overriding moral obligation to create post-persons, at least in Agar's view, it follows that we should take steps to avoid doing so, and thus prevent the highly likely and highly undesirable consequences that would follow.

This leaves unclear exactly what situations might justify sacrificing the interests of mere persons in favour of those of post-persons. One possibility is that the interests of mere persons might be permissibly sacrificed in supreme emergencies. For example, McMahan observes that most of us would countenance the killing of an innocent person if it "were necessary as a means of preventing a very large number of other innocent people from being intentionally killed by others" (McMahan 2009, 598). That being so, it might be morally acceptable to sacrifice an innocent person (but not an innocent post-person) to protect a smaller number of post-persons (McMahan 2009, 601). Agar suggests that post-persons would also be prepared to sacrifice, and perhaps be morally justified in sacrificing, mere persons when *supreme opportunities* arise: that is, situations where sufficiently great benefits can be most effectively secured by sacrificing morally significant beings, as happens now in much biomedical research on non-human animals that is aimed at improving human welfare (Agar 2014, 184–189).

Thus, Agar assumes that post-persons would sometimes be justified in sacrificing the interests of mere persons. I've cast doubt on both the existence of objective moral justification and the possibility of post-persons. But those doubts leave open the possibility that radically enhanced posthumans will *regard* themselves either as post-persons (with a higher moral status) or simply as beings with far weightier interests than ours. The latter might be rationalized by the belief that unenhanced or moderately enhanced human persons cannot feel anything like the heights and depths of experience that the posthumans enjoy and endure. If there's a sufficient cognitive gulf between their abilities and those of historical and present-day *Homo sapiens*, this belief might be true. They might live, experience the world, contrive schemes, and engage in strategies that we'd find baffling. By way of comparison, even if we recognize gorillas as persons we are unlikely to give priority to their needs over those of human beings, and we can provide a more-or-less plausible rationale for this.

Alternatively, the posthumans' approach might be based on sheer feelings of solidarity among themselves to the exclusion of everyone else – such as *us*. They might treat us with a certain amount of kindness, but we'd find our interests radically deprioritized. Conversely, if unenhanced and moderately enhanced humans had the upper hand, perhaps by sheer numbers or perhaps through residual control of important institutions, the result might be that radically enhanced posthumans who revealed such attitudes would be put down without mercy.

Once a cognitive gulf opens up between two groups, it is difficult to close it with a common moral system accepted on both sides. If conflict arises, the two sides might operate with different, and conflicting, moral theories, neither of which is objectively binding on all concerned. As Agar suggests in *Humanity's End*, superintelligent posthumans on the far side of a cognitive gulf might employ social contract theorizing, in which case they might view more ordinary humans like us as less than full partners in a social contract, or as poor candidates to enter a social contract with them. If the gulf is vast, "posthuman social contract theorists may conclude, correctly [by the lights of their theory], that humans have no moral value at all" (Agar 2010, 163).

Alternatively, superintelligent utilitarians – or moral consequentialists more generally – might reason that their own welfare counts for more than that of the humans stranded on the other side of the cognitive gulf. Like their present-day human equivalents they would pay most attention to those beings deemed to have the highest capacity for rationality, self-awareness, and richness of desires and experiences – which would doubtless mean themselves (Agar 2010, chap. 8). But once again, ordinary human persons could not be compelled by rational inquiry (as opposed to superior force) to recognize posthumans as post-persons or to embrace their moral theories. If conflict arose, they might develop an ethic based on ideas of solidarity and liberation, and they might view the posthumans simply as enemies to be defeated.

6. Tying the loose ends

In this chapter, I distinguished moral enhancement from moral status enhancement, then focused on the implications of the latter. Here, technology is employed to create beings who, actually or allegedly, deserve a higher degree of moral consideration than unenhanced, or moderately enhanced, human beings, or than any other creatures that we might regard as persons. If moral status enhancement is possible at all, it seemingly depends on radical cognitive enhancement, perhaps accompanied by some form of radical moral enhancement.

I have cast doubt on the possibility of moral status enhancement, and indeed on the classic theory of moral status that supports it. Nonetheless, sufficiently advanced posthumans might regard themselves as post-persons or simply as beings with interests significantly weightier than those of unenhanced or less enhanced humans. They might, therefore, claim superior moral rights and/or grant themselves superior legal rights. We can imagine plausible scenarios in which this leads to social conflict or to social relations between groups of dominance and subordination. These outcomes, I take it, are plainly undesirable, and they suggest pessimism about the viability of mixed human–posthuman societies.

Although I've disputed much of Nicholas Agar's analysis, I agree that this issue poses a serious problem for proponents of radical cognitive enhancement. This requires more thought, and it leads to a question for Chapter Ten: "Can't we all get along?" Short of suppressing radical enhancement – especially radical cognitive enhancement – completely, is there a path to a better and more harmonious future?

Chapter Ten : Can't We All Get Along?

1. Some pessimism

In some ways, the preceding chapters are pessimistic. I've almost ruled out radical enhancement of human physical capacities such as strength and athleticism, interpreting them as positional goods with no – or little – intrinsic value for those who receive them and little value for society as a whole. For some important professions, there might be good reasons for *moderate* enhancement of strength and other physical characteristics. But there seems to be no good reason, other than attempts to obtain a positional advantage, for radical enhancements that would bestow physical abilities dramatically greater than those of elite athletes.

In Chapter Seven, I acknowledged a potentially serious impact from radical life extension at a time when humanity already faces a Malthusian crisis. We could expect a significant effect on current demographic projections if radical longevity became commonplace on a global level any time before deep into next century. Even if it were confined to a relatively elite group such as the wealthiest 10 or 20 per cent of the world's population, the impact could worsen the situation for everyone else, causing a divergence of interests between rival populations in richer and poorer societies and/or between richer and poorer sub-populations within the same societies.

In any event, developing workable forms of radical life extension will be difficult. We have some ideas and some provocative data from studies on non-human animals, and we have some suggestive data from formal and informal research on human beings. So far, however, we've had no clear successes in slowing or halting, much less reversing, human senescence. Even on an optimistic interpretation of the current state of play with life-extension research, it might be many decades before we have reliable therapies for extending the maximum span of human life by even a small number of years.

Furthermore, efforts to cure senescence, and to produce lives extending for centuries or more, will need to include efforts to expand human memory and enhance ageing brains. Otherwise, negligibly senescent people could lose touch with their earlier lives and lose important kinds of fluidity and openness in their thinking. Although I have largely bracketed them off from this study, ethical requirements in research are also likely to hamper at least some research into radical life extension (Agar 2014, chap. 6).

None of this is an argument against initially moderate efforts at retarding senescence and prolonging life. While these also face problems from the complexity of the issues and confounding factors in research, they will doubtless continue. Successful research of this kind will produce straightforward benefits for individuals and societies. Nonetheless, radical therapies that could produce lives extending for hundreds or thousands of years appear to be far away.

Cognitive enhancement might take many forms, from the mild benefits of drinking coffee to amplified human intellect on such as scale as to create a gulf between unenhanced or moderately enhanced humans and people whom we'd intuitively regard as posthuman. Here, too, I've shown some pessimism.

At present, we have little idea how to produce even moderate forms of cognitive enhancement. Genetic editing of human DNA to raise the potential for intelligence is a poor candidate as a technology of radical cognitive enhancement, even if we identify relevant genetic contributions for levels of intelligence within the normal human range. On even the most enthusiastic guesses, genetics offers no reasonable prospect of creating intellects significantly beyond historically observed levels. Perhaps we can soon embed cybernetic devices in our skulls, interfacing directly with our brains and nervous systems, but even this does little more than bond us more closely with a tool that we could already use. Some more advanced kind of cybernetic enhancement might open a cognitive gulf and justify fears of morally greedy post-persons,[43] separate human and posthuman social contracts, and other such outcomes, but this is not currently on the horizon.

43 See Horta 2017, 908–909.

In fact, we have no real conception of what radical cognitive enhancement would be like. Anyone who has used, or seen, cranes and forklifts, has a sketchy idea of what it would mean to lift huge weights with our own muscles (perhaps the feeling of effort would not be so different from what we experience now with considerably smaller weights).[44] We might also have an idea of what it would be like to retain youthful vigour far beyond our thirties, or even beyond the robust middle years of our forties and fifties. If we've reached adulthood, we're either enjoying that kind of vigour now or we can remember it from the past. By contrast, radical cognitive enhancement is an insurmountable challenge even for our imaginations.

None of this justifies rejection of more moderate – if significant – kinds of cognitive enhancement, or of incremental research and social implementation that could lead to superintelligent people emerging over time. Nonetheless, this is an area for caution. Allowing a cognitive gulf to develop between sub-populations would be socially risky, and public policy should be set against it.

Each imagined form of enhancement – I have not dealt with them all – has its own attractions and problems. However, no proposed form of *radical* enhancement of which I'm aware is straightforward, either scientifically or in its possible impact. As I've maintained throughout this book, radical enhancement will not leave our lives much as they were except for a few improvements. It will, instead, have far-reaching and unpredictable effects.

These notes of pessimism and caution are not my rationalizations of bioconservative instincts or an apologist approach to mortality. I don't have an agenda of preserving human capacities, human nature, and the human species and genome in a pre-technological state. On the contrary, I see much merit in enhancing some human capacities, and specifically in working over time to eliminate or reduce senescence as a limiting factor in the human condition. If anything, I'm intrigued and excited by the idea of a future posthuman civilization in which very capable people live very long lives. This is not something that I view with revulsion or resignation. Realistically, however, it

44 Obtaining enormous strength would probably be a transformative experience, but we can imagine in a basic way what using our enhanced strength might feel like.

cannot be achieved within a single lifetime or (probably) a single century. It is building a cathedral over multiple generations.

2. Some optimism and some realism

Although caution is appropriate when dealing with anything as dramatic as radical enhancement, some specific concerns can be deflated. Nicholas Agar has made the point that we should make transformative choices using our current values.[45] It would be mistaken to choose, for example, the option of radical longevity based on the values we expect to form *after* our lives have been extended. By way of comparison, we would not choose to become Cybermen even if we thought that we'd endorse the decision afterwards. We should – we actually have no alternative – make jungle-or-ship choices using the value systems that we have now.

While this is correct as far as it goes, and conforms to standard decision theory, our current value systems are complex, multi-leveled, and not entirely transparent even to ourselves. They might include an emotional attachment to remaining much as we are, and as humans have been historically, but they might also include obtaining greater understanding and agency (in the sense of capability or power) and overcoming our human limitations. We might place considerable *dis*value on those limitations and place no value in particular on being *Homo sapiens* in the sense defined by the biological species concept – or on being human, rather than posthuman, in any more informal and intuitive sense. We might, moreover, value experiences that could change us, even if they changed our values in the process, although there might be certain *specific* values that we would never want to take up. Although we should not, and cannot, be motivated by values that we don't currently possess, we can place a positive or negative value on how happy or miserable we expect to be after undergoing a transformative experience.

All of this suggests that the injunction to rely on our current values sounds simpler than it really is for the transformative choices that confront us in our lives. On the one hand, we have no alternative. On the other hand, it's complicated. Our current values might preclude a choice to become Cy-

45 He would want us to use species-relative universal human values.

bermen, but they don't necessarily preclude other transformative choices, such as to have a child, to become a doctor, to join the military, or to undergo therapies for radical life extension. Some of us might rationally choose radical longevity even if we thought it would have a downside such as increasing our fear of death. Here, the choice depends on the respective value systems of individuals. Nicholas Agar and Aubrey de Grey might both be rational in their contrasting choices, given that they start with differing values.

The good news is that we have no evidence that radical longevity actually would be accompanied by a heightened fear of death. If anything, the evidence seems to run the other way: it suggests that death might seem more remote to people who've obtained radical longevity, and it might be less a factor in their thinking. Likewise, we should not be persuaded by the argument that we would, after a couple of hundred years, lose contact with our original selves and not survive radical life extension. As a caveat, however, this assumes that progress in overcoming senescence includes progress in enhancing ageing minds.

In all, at the personal level, radical longevity does not appear to be a "poisoned chalice" (Bhattacharya and Simpson 2014, 234), provided that it is developed carefully over time, perhaps in installments as new anti-ageing therapies come on stream. Agar writes of radical longevity: "We may persist, but only with existences that we properly view as impoverished" (Agar 2010, 198) – but that is too pessimistic. It might become true if new technology is developed hastily or handled poorly. But just as realistically, human beings might flourish in wonderful, scarcely conceivable ways once liberated from the slow-drip poison of senescence and the ever-present shadow of mortality that falls over human lives.

Radical enhancement could cause problems, but in the main these are not about undesirable outcomes for individuals. More worrying are the social and environmental issues that might arise, especially if the interests of different groups diverge within mixed societies of people who have or have not had the opportunity and motivation to employ radical enhancement technologies. These mixed societies might be transitional, existing for a period between our current situation and a time when radical enhancement is widespread and a different form of civilization is in evidence. If, however, mixed human–posthuman societies are conflict-ridden or marked by gross inequalities of power, that could swerve our future in dangerous directions.

3. Cyborg citizenship

In his important manifesto *Citizen Cyborg*, James Hughes introduces the idea of democratic transhumanism as a combination of social democratic and transhumanist ideologies. At a more precise level of analysis, it is a combination of culturally progressive, economically egalitarian, and technologically optimistic viewpoints (Hughes 2004, 73). For Hughes, democratic transhumanism is one position among eight theoretical combinations of political ideas, represented in three dimensions as the eight corners of a cube.

Some combinations are more plausible than others as real-world political positions, and Hughes largely confines himself to analyzing just four: generic right-wing and left-wing versions of transhumanism (or at least forms of optimism about technology); and generic right-wing and left-wing varieties of what he calls bioLuddism. He does not expect any culturally conservative form of transhumanism to become "a very popular variety" (Hughes 2004, 181). This seems right, as transhumanist ideas are inherently likely to shake up traditions and established values. While that is compatible with right-wing forms of libertarianism, it runs contrary to culturally conservative sentiment. To complicate matters further, all eight theoretical combinations are ideal types, since an individual might occupy any of an indefinite range of points on each of the three axes that Hughes explains.

In *Humanity's End*, Agar examines the idea of democratic transhumanism in a way that is slightly misleading, though ultimately consistent with Hughes's vision of the human future. Agar's explanation is as follows:

> The citizens of societies organized according to the principles of democratic transhumanism will understand that the vast differentials in power between the unenhanced and the radically enhanced have no bearing on their moral worth. They will view a citizen's IQ or life expectancy as irrelevant to his or her political standing. Hughes thinks that democratic transhumanism can ensure a harmonious future for societies that comprise individuals who are making the transition from humanity to posthumanity at varying speeds, or not at all. (Agar 2010, 11)

Although Hughes does not *define* democratic transhumanism, in this way, it does summarize a moral and political stance that he expects democratic transhumanists to adopt. Agar's account is best understood as a summary of Hughes's related concept of cyborg citizenship: the idea of citizenship based on personhood rather than on species-membership (Hughes 2004, 79). Nothing of substance depends on this terminological point, but it is best to get the relevant concepts and terminology clear.

For both Hughes and Agar, the issue of substance is whether ideas such as cyborg citzenship could provide – or significantly inform – the moral and political principles of a future mixed society of humans and posthumans, keeping in mind the dismal history of encounters between human beings empowered by different levels of technology (Agar 2010, 151). That history includes endless examples of conquest, dispossession, enslavement, and worse. In response, Hughes proposes social arrangements based on the classic three-level theory of moral status that we examined in Chapter Nine. For Hughes, citizenship should be founded on the Lockean conception of personhood, rather than on membership in a particular species such as *Homo sapiens* (or on something even more parochial, such as a particular culture or ancestry). Cyborg citizenship would supersede existing concepts of universal rights based on our common humanity. Hughes follows the classic theory of moral status in treating personhood as a threshold property: you either have it or you don't, and it doesn't come in degrees.

Unlike many philosophers and bioethicists, Hughes is willing to employ the classic theory rigorously. He recognizes some existing non-human animals, including chimpanzees, gorillas, and dolphins, as Lockean persons, and hence eligible for citizenship rights. Accordingly, he envisages a democratic transhumanist society as extending "upward" to include radically posthuman citizens and also "downward" to include chimpanzees and some other notably intelligent mammals. It does not follow that chimpanzees would possess all the same legal rights as adult humans, but they would have a considerable panoply of rights and protections similar to those assigned to human children. As an additional, and important, feature of his ideas, Hughes proposes a right for less capable persons to request, and be given, assistance in upgrading their capacities. We might even provide this assistance without request to those persons (notably apes) that are unable to request it. As Agar expresses this point, "The early adopters of radical enhancement will acknowledge a duty to help the rest of us to follow them" (Agar 2010, 156).

All of this underlines that Hughes is imagining a *transitional* society. On his account, the coexistence of beings with greatly different capacities would not be permanent. Meanwhile, the concept of cyborg citizenship would smooth the transition and sustain harmony even in radically diverse societies with citizens whose cognitive powers span a wide range. Not unexpectedly, Agar responds with scepticism. He predicts that cognitively sophisticated posthumans would, in practice, "view humans as morally required to defer to them, to permit our interests to be sacrificed to promote theirs" (Agar 2010, 152). If so, widespread promulgation of ideas such as cyborg citizenship would have limited efficacy. As we saw in Chapter Nine, Agar makes a plausible case that posthumans on the other side from us of a cognitive gulf would develop their own moral theories in which their interests are privileged over ours.

There is, in fact, an even deeper problem with cyborg citizenship, though it is not raised by Agar. Cyborg citizenship relies on the classic three-level theory of moral status, but at best this theory is not straightforwardly true. It does latch on to objectively real characteristics such as sentience and self-consciousness, but insofar as it prioritizes these characteristics and tells us how to respond to them, it is not objectively authoritative. In practice, it is trumped by human beings' natural responsiveness to each other and by our deep concern for the welfare of very young humans (even before they obtain the levels of self-consciousness that we believe are found in adult apes and cetaceans).

Hughes would doubtless view this as morally impermissible speciesism warping our thinking, but it seems to me that it is an ineradicable element of human moral systems. These systems do not exist to promote the welfare of all creatures, or all rational creatures, but (in essence) to promote cooperation and peace within human societies (Blackford 2016 chap. 2; Blackford 2019). Increasingly, this requires cooperation *among* human societies, but it does not require cooperation with, or solicitude toward, chimpanzees, dolphins, and other non-human creatures that are probably, or arguably, Lockean persons. That is a repurposing of morality, perhaps a by-product of increasing technological power, which means that most non-human animals are no longer credible threats or competitors to human groups, and increasingly they are viewed with sympathy. Greater respect for apes and cetaceans fits with my own values, but I can't demonstrate that it is obligatory to people – human or posthuman – whose value systems are different from mine.

Posthumans *might* voluntarily sign up to support the idea of cyborg citizenship, but this is not rationally demanded of them. If posthumans live and think on the other side of a cognitive gulf, I don't like the chances that they'll view us, on our side of the gulf, as their moral equals.

4. Synchronic and diachronic messiness

Having laid out his concerns about whether democratic transhumanist or cyborg citizenship principles would be adopted in practice by a mixed society of ordinary humans and cognitively superior posthumans, Agar reports Hughes's response:

> If posthumans are clearly distinguishable in terms of their behavior and appearance from humans, then, he grants, there could be a "them and us" mentality. But Hughes thinks that the reality will be considerably messier than that, and that this messiness will be good for humans. (Agar 2010, 172)

The idea, then, is that we should not expect two plainly visible sub-populations or castes, but rather a wide range of cross-cutting identities and alliances not greatly different from what is seen in current liberal democracies. At any particular time, there will be a "messy" situation in a mixed human–posthuman society. For example, gay posthumans will be more similar to gay humans – at least in one important respect – than to straight posthumans. Differences between humans and posthumans will be partly submerged by other differences, such as those of race, sexuality, religion, and political ideology – as well as numerous individual styles, idiosyncracies, and so on – which will lessen any straightforward tendency for society to split into factions of posthumans and "mere" humans. We can think of this as synchronic messiness: a messy situation across a society at a particular time.

For Agar, immense cognitive differences are likely to overwhelm similarities in other dimensions. By analogy, he suggests, we'd expect a gay man who was born in early August to feel little commonality with a male macaque that shares his star sign, has similar coloured hair to his, and frequently mounts other males of its species. In such a case, the cognitive gulf is too large, and even if the man keeps the monkey as a pet and feels fond of it (perhaps partly because of their points of resemblance), he will not view it as entitled to the same moral consideration as human beings. Agar's fear is that posthumans in a mixed society would regard us much as we regard the macaque, and they might see any points of resemblance between a particular posthuman and a particular human as trivial. Thus, "There's a good chance that posthumans whose sexual practices are the nearest equivalent to gay may experience a similar indifference toward homosexual humans" (Agar 2010, 172–173).

It's difficult, however, to assess whether there's a good chance or a not-so-good chance. The situation is more uncertain. We do know that the similarities between the gay man and his macaque in the original version of the thought experiment really were trivial in the sense that the man and the monkey stood on opposite sides of a dramatic divide in their inner experience. As far as we know, macaques lack any sophisticated sense of their own future, do not make plans or harbour ambitions, and have little or no ability to make decisions on the basis of reasons, or to query their own momentary impulses. They are not, as far as we can tell, capable of emotions such as deep sadness and grief, and nor can they feel the elation of having their plans work out for the best after conscientious striving.

If we turned out to be wrong about macaques – if they were more like us in these respects – we might think twice about giving them so little moral consideration. By comparison, cognitively sophisticated posthumans will know much about how we think and feel and experience the world. Although they might see their interests as more weighty than ours, I'd want to know much more about them before guessing the odds of whether they'd treat us as lightly as we treat macaques. Perhaps they would regard us not as disposable creatures, as we perhaps too willingly think of non-human animals, but as intelligent beings who deserve considerable deference and protection, but who are not competent for some tasks that the posthumans find routine. This is a less dreadful possibility than some others postulated in the academic literature on radical enhancement, but it assigns an unsatisfactory place for unenhanced and moderately enhanced humans in a mixed, transitional society.[46]

This is, unfortunately, a situation where philosophical thinking runs out of power. Hughes and Agar are engaged in empirical speculations about future societies that prominently include beings unlike any that we have ever encountered. Getting drawn into the game of speculations might be unavoidable, but we should be clear that nothing is truly knowable in this territory and that the most we can do is make intelligent, but under-informed, guesses. My best guess is that the idea of cyborg citizenship and the likelihood of some synchronic messiness will be helpful for averting bad outcomes, but they'll not be sufficient to produce good outcomes in a society with a wide cognitive gulf.

We can view cyborg citizenship not as a component of an objectively authoritative morality – or a universal human morality – but as a proposed political principle that could assist us now and in any future transitional human–posthuman society. Viewed in this way, the idea can usefully be promulgated in books such as *Citizen Cyborg*, taught to children, assumed in works of fiction written by individuals who endorse it, and so on, until it joins other revered political principles such as freedom of speech, freedom of religion, and due process for individuals accused of crimes. Agar himself appears to find merit in such ideas as democratic transhumanism and cyborg citizenship, while being understandably sceptical of how they'd fare in a mixed society.

At this stage of the evolution of industrially advanced societies, cyborg citizenship is likely to attract adherents. Since it takes personhood so seriously, it entails investigating the claims to personhood of various non-human animals and giving particular respect, including special legal protections, to whichever creatures appear to cross the threshold. For many of us, with our current values, that is attractive in itself. This "downward" respect for personhood might also create a helpful precedent for societies where some people are cognitively enhanced and some are not, or where some are more en-

46 Compare the relatively benign two-tiered society of unenhanced humans and enhanced super-cooperators described by Allen Buchanan in his 2009 article on moral status and human enhancement (Buchanan 2009, 372–381).

hanced than others. But I'm still not convinced that a principle of cyborg citizenship would withstand the pressures of a society with a true gulf between the abilities of *radically* enhanced citizens who stand to ordinary adult human beings much as the latter stand to gorillas and chimpanzees. The contrast in competence would be too great, and there is too much scope to devise alternative moral theories and political principles that radically enhanced citizens would find plausible – and which would give their interests priority. Although this is speculative, it seems that cyborg citizenship can be only part of the story of how to avoid societies with stratification or intergroup conflict.

Likewise, Hughes's point about the messiness of transitional human–posthuman societies is important, but it is still no guarantee of social harmony, and an acceptable level of equality, once we reach societies marked – or scarred – by cognitive gulfs. An emphasis on toleration and pluralism has the benefit of allowing people with widely different characteristics, values, understandings of the world, and so on, to coexist and flourish in the same society, and it might help to smooth over further differences relating to different modes, types, and levels of enhancement. This suggests a role for traditional liberal tolerance and social pluralism. But again, the resulting benign messiness across a society at any particular time is probably not enough. High levels of toleration of difference can take us a long way, but it's doubtful how far they can assist once the cognitive competence of some citizens incomparably exceeds that of ordinary adult humans.

In practice, there is likely to be diachronic, as well as synchronic, messiness. That is, we can expect to see a varied and unpredictable set of snapshots of what any specific society looks like over time as enhancement technologies are developed and disseminated. If this is handled well, it will mean that immense gulfs in cognitive and other capabilities do not arise or are isolated and manageable. Ideally, successive waves of technology will be taken up almost universally, with sufficient breathing space for individuals and societies to adapt. Diachronic messiness could provide a large part of the solution that we need, but it flies in the face of many ambitions for near-term, apocalyptic change.

5. Making our peace with the future

Early in the text of *Truly Human Enhancement*, Agar states, "It would be a serious mistake to just assume a scenario in which enhancements arrive in small increments with plenty of time for us to adjust between each installment" (Agar 2014, 5). Here he points to a real concern, so allow me to underline it.

Consider a scenario of abrupt arrival of radical enhancement technologies to existing societies under current conditions. Imagine that the package of enhancements includes highly effective life extension plus cognitive enhancement at a level that would open a gulf in competence between those who do and don't gain access. If large numbers of people start living to unprecedented ages, this soon adds to Malthusian pressures. Likewise, enhancement of a kind that opens a cognitive gulf opens up the prospect of an oppressive or conflict-ridden two-tier society.

Yet, such a scenario is not based on a scientific consensus that particular technologies can be developed and made available in a particular timeframe. We should not underestimate the power of technological change over generations, or we'll resemble Edgar Allan Poe imagining long-range travel by balloon one thousand years after his short story "Mellonta Tauta" (1848) was published. But likewise, many people not many decades ago expected undersea cities and colonies on the moon by now. These things cannot be predicted with much accuracy. All we know for the moment is that the most exciting, yet troubling, technologies face many hurdles before they can be developed, tested, and widely disseminated and implemented.

The hype about imminent and abrupt introduction of radical enhancement technologies does not come from reliable knowledge of what level of innovation is possible at what pace. It stems, rather, from the theorizing of a small number of high-profile advocates of sweeping change to the human condition and to human beings. Among these advocates are Aubrey de Grey and Ray Kurzweil, who are clearly impressive individuals. Perhaps history will validate their ideas and predictions, but at this stage of the curve of progress we simply don't know. It's most reasonable to expect that effective technologies for radical enhancement will be preceded by waves of partial and/or more moderate equivalents. Assumptions to the contrary are based on extreme optimism (or pessimism, if that's how it seems from within your value system).

In the final pages of *Humanity's End*, Agar contrasts the gradual changes to humanity envisaged by Julian Huxley in the 1950s with the more abrupt changes imagined by some of today's proponents of radical enhancement:

> Over many generations, [the eugenic practices imagined by Huxley] could lead to beings that are so different from us that they are not properly considered human. But the gradual nature of this transformation wouldn't prevent us from relating to our former selves, our children, and our fellow citizens in ways compatible with our human values. We and they will be human. The technological transformations advocated by early twenty-first century defenders of radical enhancement differ in being abrupt. (Agar 2010, 197)

Unlike Condorcet and Godwin in the 1790s, we've experienced the rise of powerful technologies that give us the ability to shape the planet and to some extent ourselves. Among many other developments over the past century, we've devised nuclear weapons and reactors, made enormous strides in robotics and information technology, revolutionized transport and communications, commenced an era of space flight and exploration of our solar system, discovered DNA as the molecular basis of life, made impressive advances in genetic technology, and taken health care to an altogether new level of sophistication. It is reasonable to expect continuing convergence of these technologies, and it's natural to speculate about perfected, superlative versions of them. In addition, the turn of a new millennium has compelling symbolic force, even if acting only unconsciously on the minds of rational people who understand the arbitrariness of the Western calendar. All of this primes us to expect the most extreme forms of technological power in our own lifetimes, including the power to fulfil our dreams. Chief among those, of course, is the ancient dream of defeating death.

The odds are against that dream abruptly coming true, and the same applies to many other dreams of personal power. If we advocate enhancement technologies while focused on their benefits for us, as individuals, we're in for disappointment. The benefits will (probably) be incremental and limited. As a corollary, fears of apocalyptic change are (probably) overblown. It's wisest, I submit, to dial back our selfishness and to think more impersonally about humanity's future.

6. How to build a cathedral

The cathedrals of the Middle Ages were constructed over generations, sometimes taking centuries from planning to completion. In each case, the idea was to create something grand and lasting, and, by the consensus of the time, undeniably worthwhile – a monument of faith and a visible, tangible representation of God's glory. Today's secular thinkers might conclude that some of this was misguided, but we can still appreciate these magnificent buildings and mourn their destruction, as with the ruination of Coventry Cathedral by German bombing during World War II[47] and the extensive fire damage to Notre-Dame de Paris in 2019. The great phases of cathedral building have passed, but even today work continues on the Basílica de la Sagrada Família in Barcelona, which commenced construction in the 1880s. When this is finished, it will be breathtaking.

Building great cathedrals required preparation, diligence, sacrifice, and time. It offered rewards, some perhaps illusory but others undeniable in the form of cultural participation and contributing to something splendid. There was, of course, often exploitation and corruption, but trust in the process led to impressive outcomes. This gives an answer to Steve Fuller's question: "So, do you still believe in transhumanism even if it is unlikely that you will personally benefit from it?" (Fuller 2020, 202). Whether or not we call ourselves transhumanists or "believe in transhumanism", we can contribute to a human future that is greater than ourselves. This need not mean researching genetic enhancement or life-extension therapies or brain–computer interfaces. Each of these fields might have its value, but there are other much-needed contributions to a new civilization. Think of this as building a great cathedral. It will take time, it calls for thought and planning, and it requires the right civic and material conditions. Right now, we're just beginning.

Recall Condorcet's assurance that those who work for "the progress of reason and the defence of liberty" join "the eternal chain of human destiny" (Condorcet 1979 [1795], 201). The word "destiny" is too strong, since the

47 This medieval church was not given cathedral status until 1918, but it was, prior to its destruction in 1940, one of the most impressive religious buildings in the United Kingdom. I can report from my own visit that its ruins are sobering and heartbreaking.

future defies prediction, good outcomes require choices and work, and even then nothing is destined or guaranteed. But we can take steps to establish societies and a global order that are better prepared for a radically enhanced future citizenry. The lesson is not that some technologies or some research should be suppressed, but we can seek ways to skew the probabilities in our favour – to raise the chance of good outcomes and to lower the risk of things going off the rails.

Exactly what should be done requires discussion beyond this book. For a start, I suggest that we – as individuals and political participants – encourage moderate enhancement technologies and their wide dissemination. The path to radical enhancement lies through moderate enhancement, and the path to social harmony lies through widespread, broadly equal access. The latter, in turn, requires more economically equal societies. Similarly, we should support social pluralism in all its dimensions. Most obviously, perhaps, we need to make progress in solving existing large-scale problems such as those posed by global warming. These are all broad notions, but they can be elaborated indefinitely with detailed data and tentative plans. Moderate enhancement, more equal societies, and social pluralism are good things from the perspective of my value system, and perhaps yours. They are worth pursuing in their own right. So much the better if they help smooth the way to a new form of civilization.

7. The human in the posthuman

In *Humanity's End*, Agar accepts that *Homo sapiens* will eventually become something different. Barring "some global or galactic catastrophe", we'll eventually "evolve into beings who are not properly considered human" (Agar 2010, 198). Likewise, in *Truly Human Enhancement* he states, "At some point, should our lineage not go extinct, we are almost certain to be replaced by beings different from us and with values different from our own" (Agar 2014, 198). His preference, however, is that any such future be distant, and that we retain what he understands as our humanity at least for the next few hundred years.

Despite his official commitment to the biological species concept, Agar often seems to have in mind a more intuitive idea of humanity and its essential values and characteristics. Even so, it is difficult to be sure exactly what he wants to preserve. From my discussion in earlier chapters, I conclude that the following, at least, are *not* under serious threat:

- Appreciation of virtues and relationships such as cooperation, friendship, moderation, fairness, and courage.
- Pleasure in developing and displaying physical skills.
- The ability to put fears of death largely out of our minds and to engage in risky activities such as mountaineering and car racing (or simply driving on ordinary roads).
- The urge to understand the world through scientific and other means.
- Our attachments to our own societies, and our hopes for their survival and success.
- Feelings of love and care for infants and children, especially our own offspring.

These might be plausible candidates for universal human values, but many other values currently or recently held by numerous human beings are merely those of particular societies under concrete historical conditions. *Those* can change as conditions change, as has happened in recent history with attitudes to war and sexuality.

That said, I agree with Agar that any transition to an intuitively posthuman civilization could, and indeed should, take place over multiple generations – perhaps not as long as a few hundred years, as he suggests, but not abruptly. For Agar, what he interprets as the end of humanity should be acknowledged with resignation. He views it as like our own eventual deaths: something to accept as inevitable without wanting to hasten the process. But this is the wrong comparison. For the individual, death is the loss of everything. It is the loss, most notably, of consciousness, but also of each person's unique "combination of qualities, defects, and experiences" – leaving behind no more than "comparatively brief narrative outlines of our lives" (Agar 2010, 198).

Notwithstanding many mythological narratives and religious promises, death is not our transformation into beings with radically improved understanding and agency. By contrast, if humanity, as we understand it scientifically and intuitively, evolves through a process of change that culminates in a posthuman civilization, this is not the *end* of humanity or of "the human story" (Agar 2014, 200). It is metamorphosis and empowerment, not cyber-conversion or Borg assimilation or any other fate as bad as death itself. Managing the process in a few years, or within one lifetime, might be impossible to achieve, and unwise to attempt – and so I have argued – but if we prepare well, our successors will emerge from *us*. Their nature will be path-dependent on their human origin, and they'll continue our story, perhaps carrying it far beyond this planet where we, and ultimately they, evolved from apelike ancestors.

On this account, we're early in the human story. Our prehistoric past was a long prologue, our civilizations to date an early phase of the action, the great transition ahead of us a bridge to something much larger still. In one sense that's a blow to our egos, for grander things lie in our future than in the past that formed us. But it's an exciting thought. If it's true, the human story is just beginning.

Bibliography

Agar, Nicholas (2004). *Liberal Eugenics: In Defence of Human Enhancement.* Oxford: Blackwell.

Agar, Nicholas (2010). *Humanity's End: Why We Should Reject Radical Enhancement.* Cambridge, MA: MIT Press.

Agar, Nicholas (2014). *Truly Human Enhancement: A Philosophical Defense of Limits.* Cambridge, MA: MIT Press.

Avery, John (1997). *Progress, Poverty and Population: Re-reading Condorcet, Godwin and Malthus.* London: Frank Cass.

Benford, Gregory (1986). *In Alien Flesh.* New York: Tor.

Benford, Gregory (1999). "A Hunger for the Infinite." In Robert Silverberg (ed.), *Far Horizons: All New Tales from the Greatest Worlds of Science Fiction.* London: Orbit, pp. 346–378.

Benford, Gregory (2000). *Worlds Vast and Various.* New York: EOS.

Bernal, J. D. (1970). *The World, the Flesh, and the Devil: An Inquiry into the Future of the Three Enemies of the Rational Soul.* London: Jonathan Cape. (Orig. pub. 1929.)

Bhattacharya, Aveek, and Robert Mark Simpson (2014). "Life in Overabundance: Agar on Life-Extension and the Fear of Death." *Ethical Theory and Moral Practice* 17: 223–236.

Blackford, Russell (2007). "Differing Vulnerabilities: The Moral Significance of Lockean Personhood." *American Journal of Bioethics* 7(1): 70–71.

Blackford, Russell (2014a). *Humanity Enhanced: Genetic Choice and the Challenge for Liberal Democracies.* Cambridge, MA: MIT Press.

Blackford, Russell (2014b). "Introduction II: Bring on the Machines." In Russell Blackford and Damien Broderick (eds.), *Intelligence Unbound: The Future of Uploaded and Machine Minds.* Chichester: Wiley-Blackwell, pp. 11–25.

Blackford, Russell (2016). *The Mystery of Moral Authority.* Basingstoke: Palgrave Macmillan.

Blackford, Russell (2019). "After Such Knowledge – What? Living and Speaking in a World Without Objective Morality." In Richard Garner and Richard Joyce (eds.), *The End of Morality: Taking Moral Abolitionism Seriously.* New York and Oxford: Routledge, pp. 59–76.

Bostrom, Nick, and Toby Ord (2006). "The Reversal Test: Eliminating Status Quo Bias in Applied Ethics." *Ethics* 116: 656–679.

Broderick, Damien. (2002). *The Spike: How Our Lives Are Being Transformed by Rapidly Advancing Technologies.* Second paperback ed. New York: Tom Doherty Associates.

Buchanan, Allen (2009). "Moral Status and Human Enhancement." *Philosophy and Public Affairs* 37: 346–381.

Čapek, Karel (1999). *Four Plays.* Trans. and introd. Peter Majer and Cathy Porter. London: Methuen.

Carstensen, Laura L. (2006). "The Influence of a Sense of Time on Human Development." *Science* 312: 1913–1915.

Clark, Andy, and David J. Chalmers (1998). "The Extended Mind." *Analysis* 58: 7–19.

Condorcet, Antoine-Nicolas de (1979). *Sketch for a Historical Picture of the Progress of the Human Mind.* Transl. June Barraclough. Westport, CT: Greenwood Press. (Orig. pub. 1795; translation orig. pub. 1955.)

Coontz, Stephanie (2005). *Marriage, a History: From Obedience to Intimacy or How Love Conquered Marriage.* New York: Viking.

Coyne, Jerry A., and H. Allen Orr (2004). *Speciation.* Sunderland, MA: Sinauer Associates.

Davis, John K. (2018). *New Methuselahs: The Ethics of Life Extension.* Cambridge, MA: MIT Press.

Douglas, Heather E. (2009). *Science, Policy, and the Value-Free Ideal.* Pittsburgh: University of Pittsburgh Press.

Douglas, Thomas (2008). "Moral Enhancement." *Journal of Applied Philosophy* 25: 228–245.

Elliott, Carl (2003). *Better Than Well: American Medicine Meets the American Dream.* New York and London: W. W. Norton.

Fenton, Elizabeth (2010). "The Perils of Failing to Enhance: A Response to Persson and Savulescu." *Journal of Medical Ethics* 36: 148–151.

Fischer, John Martin (2006). "Epicureanism About Death and Immortality." *Journal of Ethics* 10: 355–381.

Fischer, John Martin (2020). *Death, Immortality, and Meaning in Life.* New York: Oxford University Press.

Frankfurt, Harry (1988). *The Importance of What We Care About: Philosophical Essays.* Cambridge: Cambridge University Press.

Fuller, Steve (2020). *Nietzschean Meditations: Untimely Thoughts at the Dawn of the Transhuman Era.* Basel: Schwabe Verlag.

Gidley, Jennifer M. (2017). *The Future: A Very Short Introduction.* Oxford: Oxford University Press.

Glannon, Walter (2002). "Identity, Prudential Concern, and Extended Lives." *Bioethics* 16: 266–283.

Godwin, William (1793). *An Enquiry Concerning Political Justice, and its Influence on General Virtue and Happiness.* London: G. G. and J. Robinson.

Greene, Joshua (2013). *Moral Tribes: Emotion, Reason, and the Gap between Us and Them.* New York: Penguin.

Gruman, Gerald G. (1977). *A History of Ideas about the Prolongation of Life: The Evolution of Prolongevity Hypotheses to 1800.* New York: Arno Press. (Orig. pub. 1966.)

Harris, John (2011). "Moral Enhancement and Freedom." *Bioethics* 25: 102–111.

Haycock, David Boyd (2008). *Mortal Coil: A Short History of Living Longer.* New Haven and London: Yale University Press.

Hitchens, Christopher (2007). *God Is Not Great: How Religion Poisons Everything.* New York: Twelve Books.

Horta, Oscar (2017). "Why the Concept of Moral Status Should be Abandoned." *Ethical Theory and Moral Practice* 20: 899–910.

Hughes, James (2004). *Citizen Cyborg: Why Democratic Societies Must Respond to the Redesigned Human of the Future.* Cambridge, MA: Westview.

Johnston, David (2011). *A Brief History of Justice.* Chichester: Wiley-Blackwell.

Kagan, Shelly (2013). *Death.* New Haven and London: Yale University Press.

Krishnamurthy, Meena (2015). "We Can Make Rational Decisions to Have a Child: On the Grounds for Rejecting L. A. Paul's Arguments." In Sarah Hannan, Samantha Brennan, and Richard Vernon (eds.), *Permissible Progeny? The Morality of Procreation and Parenting.* New York: Oxford University Press, pp. 170–183.

Kurzweil, Ray (2005). *The Singularity is Near: When Humans Transcend Biology.* New York: Viking.

Loudon, Jane (Webb) (1994). *The Mummy! A Tale of the Twenty-Second Century.* Ann Arbor: University of Michigan Press. (Orig. pub. anonymously 1827.)

McMahan, Jeff (2009). "Cognitive Disability and Cognitive Enhancement." *Metaphilosophy* 40: 582–605.

Manzocco, Roberto (2019). *Transhumanism – Engineering the Human Condition: History, Philosophy and Current Status.* Cham: Springer.

Martens, Rhonda (2019). "Can We Make Wise Decisions to Modify Ourselves? Some Problems and Considerations." *Journal of Evolution and Technology* 29(1): 1–18.

May, Andrew (2018). *Rockets and Ray Guns: The Sci-Fi Science of the Cold War.* Cham: Springer.

Medawar, P. B. (1981). *The Uniqueness of the Individual.* New York: Dover.

More, Max, and Natasha Vita-More (eds.) (2013). *The Transhumanist Reader: Classical and Contemporary Essays on the Science, Technology, and Philosophy of the Human Future.* Chichester: Wiley-Blackwell.

Mumford, Stephen (2012). *Watching Sport: Aesthetics, Ethics and Emotion.* London and New York: Routledge.

Nebel, Jacob M. (2015). "Status Quo Bias, Rationality, and Conservatism about Value." *Ethics* 15: 449–476.

Nussbaum, Martha C. (1994). *The Therapy of Desire: Theory and Practice in Hellenistic Ethics.* Princeton: Princeton University Press.

Nussbaum, Martha, C. (2013). "The Damage of Death: Incomplete Arguments and False Consolations." In James Stacey Taylor (ed.), *The Metaphysics and Ethics of Death: New Essays.* New York: Oxford University Press, pp. 25–43.

Papineau, David (2017). *Knowing the Score: What Sports Can Teach Us About Philosophy (And What Philosophy Can Teach Us About Sports).* New York: Basic Books.

Parfit, Derek (1984). *Reasons and Persons.* Oxford: Oxford University Press.

Paul, L. A. (2014). *Transformative Experience.* Oxford: Oxford University Press.

Persson, Ingmar, and Julian Savulescu (2008). "The Perils of Cognitive Enhancement and the Urgent Imperative to Enhance the Moral Character of Humanity." *Journal of Applied Philosophy* 25: 162–177.

Persson, Ingmar, and Julian Savulescu (2011). "The Turn for Ultimate Harm: A Reply to Fenton." *Journal of Medical Ethics* 37: 441–444.

Persson, Ingmar, and Julian Savulescu (2012). *Unfit for the Future: The Need for Moral Enhancement.* Oxford: Oxford University Press.

President's Council on Bioethics (2003). *Beyond Therapy: Biotechnology and the Pursuit of Happiness.* Washington, DC: Dana Press.

Scheffler, Samuel, et al. (2013). *Death and the Afterlife.* Edited by Niko Kolodny. Oxford: Oxford University Press.

Scholes, Robert (1975). *Structural Fabulation: An Essay on Fiction of the Future.* Notre Dame: Notre Dame University Press.

Sinclair, David A., and Matthew D. LaPlante (2019). *Lifespan: Why We Age – and Why We Don't Have To.* New York: Atria Books.

Strawson, Galen (2010). *Freedom and Belief.* Rev. ed. Oxford: Oxford University Press.

Weiner, Jonathan (2011). *Long for this World: The Strange Science of Immortality.* Paperback ed. New York: HarperCollins.

Wells, H. G. (1902). *The Discovery of the Future: A Discourse Delivered to the Royal Institution on January 24, 1902.* London: T. Fisher Unwin.

Wells, H. G. (1939). *The Fate of Homo Sapiens: An Unemotional Statement of the Things that are Happening to Him Now, and of the Immediate Possibilities Confronting Him.* London: Secker and Warburg.

Williams, Bernard (1973). *Problems of the Self: Philosophical Papers 1956–1972.* London: Cambridge University Press.

Zhavoronkov, Alex (2013). *The Ageless Generation: How Advances in Biomedicine will Transform the Global Economy.* New York: Palgrave Macmillan.

Index of Names

Index of Subjects

Schwabe Verlag's signet was Johannes Petri's printer's mark. His printing workshop was established in Basel in 1488 and was the origin of today's Schwabe Verlag. The signet refers back to the beginnings of the printing press, and originated in the entourage of Hans Holbein. It illustrates a verse of Jeremiah 23:29: 'Is not my word like fire, says the Lord, and like a hammer that breaks a rock in pieces?'